Polymer Concretes

Advanced Construction Materials

Mostafa Hassani Niaki and
Morteza Ghorbanzadeh Ahangari

CRC Press is an imprint of the
Taylor & Francis Group, an **informa** business

First edition published 2023
by CRC Press
6000 Broken Sound Parkway NW, Suite 300, Boca Raton, FL 33487-2742

and by CRC Press
4 Park Square, Milton Park, Abingdon, Oxon, OX14 4RN

CRC Press is an imprint of Taylor & Francis Group, LLC

Library of Congress Cataloging-in-Publication Data
Names: Niaki, Mostafa Hassani, author. | Ahangari, Morteza Ghorbanzadeh, author.
Title: Polymer concretes: advanced construction materials / Mostafa Hassani Niaki, Morteza Ghorbanzadeh Ahangari.
Description: First edition. | Boca Raton: CRC Press, 2023. | Includes bibliographical references and index.
Identifiers: LCCN 2022033129 | ISBN 9781032352640 (hbk) | ISBN 9781032353104 (pbk) | ISBN 9781003326311 (ebk)
Subjects: LCSH: Polymer-impregnated concrete.
Classification: LCC TA443.P58 N53 2023 | DDC 620.1/36—dc23/eng/20221011
LC record available at https://lccn.loc.gov/2022033129

ISBN: 978-1-032-35264-0 (hbk)
ISBN: 978-1-032-35310-4 (pbk)
ISBN: 978-1-003-32631-1 (ebk)

DOI: 10.1201/9781003326311

Typeset in Times
by codeMantra

Contents

Foreword

Polymer concrete (PC) is widely used in different industries and in our daily life because of their diverse functionality, excellent physicomechanical properties, chemical stability, low permeability to water and aggressive media, lightweight, easy placement and ability to form complex shapes, rapid curing and setting, powerful bond to most surfaces, good long-term freeze–thaw durability, and remarkable sound and thermal insulation properties. Due to its impressive properties, PC can be considered as an advanced and multi-functional construction composite material. It has become a very active research topic in the last two decades, and its use has been growing rapidly in recent years.

This book attempts to provide a comprehensive study on PCs and cites many excellent studies that have already been done on the various aspects of these materials. This scholarly book will be useful for those who are interested in advanced construction materials and polymer composites and can be used in education and research as well as commercial purposes.

Preface

Engineering advances in novel materials in the construction industry have led to the expansion of the use of polymers and polymer composites. Attempts to develop these products for special applications have led to the manufacture of a type of polymer composite called polymer concrete (PC). PC, which is made from a combination of polymers, aggregates, and fillers, can be a very good alternative to conventional cementitious concretes in special applications. The presence of polymer in its structure makes PC have much better mechanical and chemical properties than conventional cement concretes. This book includes explanations and achievements of theoretical and experimental studies in the field of PC as an advanced construction material.

So far, much research has been done in the field of development and improvement of PC properties, which have considered these materials from different points of view. But this book attempts to provide comprehensive and classified information in this field by using previous studies.

For this purpose, after defining and stating the historical perspective of use, the applications of PC are classified and the advantages and disadvantages of their use are described. Furthermore, the materials (resin, aggregates, micro fillers, fibers, and nanofillers) and their effects on PC are systematically summarized and discussed. In the following, the fabrication method of PC is explained and the standards for testing its properties are presented. Next, various properties of PC are investigated. The characterizations addressed in this book are physical properties (density, porosity, ductility, water absorption, water resistance, flowability, and shrinkage), mechanical properties (compressive, flexural, tensile, impact, fracture, bonding and interfacial shear, fatigue, creep, damping, abrasion, and wear resistance), thermal properties (thermal expansion, thermal conductivity, elevated temperature, and thermal cycle effect, fire resistance, and flammability and hot water resistance), chemical and electrical properties as well as environmental (weathering and ultraviolet radiation) properties. Finally, a summary and outlooks for PC will be expressed.

Abbreviations

ANPs	Alumina nanoparticles
CNTs	Carbon nanotubes
CTE	Coefficient of thermal expansion
CTOD	Crack tip opening displacement
CMOD	Crack tip opening displacement
J	Creep compliance
DT	Destructive testing
G_f	Fracture energy
K_{Ic} **or** K_{If}	Fracture toughness, Critical stress intensity factor
GFRP	Glass-fiber-reinforced polymer
HDT	Heat distortion temperature
MMA	Methyl methacrylate
K_{IIc} **or** K_{IIf}	Mode II fracture toughness
MMT	Montmorillonite
MWCNTs	Multi-walled carbon nanotubes
NDT	Non-destructive testing
OPC	Ordinary Portland cement
PC	Polymer concrete
PMMA	Polymethyl methacrylate
RCSS	Rapid-cooled steel slag
SEM	Scanning electron microscopy
SCB	Semi-circular bend
SNPs	Silica nanoparticles
SENB	Single edge notch bending
k **or** λ	Thermal conductivity
UP	Unsaturated polyester
wt%	Weight fraction
XRD	X-ray diffraction

1 Introduction to Polymer Concrete

Abstract

Efforts to achieve advanced and modern materials with remarkable properties in the construction industry led to the introduction and use of polymers and polymer composites in this field. This chapter introduces polymer concrete (PC) as an advanced polymer composite material in the construction industry. Structural advantages of PC over conventional cement concretes make this material a suitable alternative in harsh and unusual conditions. In this chapter, while stating the properties of PC, the history of its use and development is expressed. Due to the unique properties of these materials, their application in various industries is increasing. In the following, these applications are categorized and expressed in different areas. Finally, the advantages and disadvantages of PC are described.

1.1 DEFINITION

Although ordinary Portland cement (OPC) concrete, due to overall specifications and relatively low price, has the most applications in the construction area, several limitations and drawbacks, such as low strength (predominantly tensile and flexural) and also failure strain, high sensitivity to low temperatures and frost, and insignificant chemical stability, led to the search for new alternatives. The above-mentioned disadvantages can be fixed by replacing the Portland cement with polymers. The fabricated polymer composites provide significant mechanical properties, excellent chemical stability, good thermal resistance, and reduced water permeability.

The introduction of polymers into the concrete industry provides polymer concrete composites, which are categorized into three main groups: polymer concrete (PC), polymer-impregnated concrete (PIC), and polymer cement concrete (PCC). The versatility of PC through the use of the polymer as a binder provides new chances for PC beyond different types of concretes. This book evaluates the features of the PC, and the other two types, PIC and PCC, are not considered.

Polymer concrete (PC) is a composite material fabricated by synthesizing polymers and fillers [1–3]. Therefore, PC does not contain Portland cement and water. Polymer concrete is also known as resin concrete. It is vital to have some cognition of the polymerization, materials specifications, and advantages and drawbacks of PCs to produce the most desirable composite for a particular need.

1.2 HISTORICAL PERSPECTIVE OF POLYMER CONCRETE

The expansion of PC began in the late 1950s when it was implemented in the United States to fabric architectural facing panels and production of synthetic marble. Later

DOI: 10.1201/9781003326311-1

the application of PC was expanded by the mid-1970s as a repairing substance for ordinary Portland cement (OPC) concrete, due to particular characteristics such as significant bond strength to cementitious and steel structures, fast curing time, and excellent mechanical properties [4].

In 1971, American Concrete Institute (ACI) 548 Committee-Polymers in Concrete was established for gathering a comprehensive database on characteristics of PCs and released "Polymers in Concrete—State-of-the-Art Report" (ACI 548R) and proceedings of ACI symposia considering the materials and manufacturing of PCs [5].

The application of precast PC structures for underground facilities started in 1975, and with the growth of the application, some standards were designed. The manufacturers implemented the published standards to fabricate more consistent products [6].

Through the 1970s and 1980s, numerous investigations on the characteristics of composites, including the PC fabrication procedure, were performed by the U.S. Federal Highway Administration, the Bureau of Reclamation, the Department of Energy, and chemical companies, and resulted in improving and expanding polymer binders implemented in PCs and prepared them for commercial productions [4].

RILEM established a Technical Committee for "Concrete Polymer Composites" (TC-105-CPC) and "Test Methods for Concrete Polymer Composites" (TC-113-CPT) and prepared different test procedures for PCs [5].

The Society of Material Science Japan (JSMS), with the aim of the synthetic resins for concrete, conducted significant research toward the development of PC materials and released design recommendations for polyester concrete structures and also instruction for mix design. The United States, the United Kingdom, Japan, Germany, and the erstwhile Soviet Union were the countries that performed standardization studies on numerous test procedures, usage, and implemented PC composites [5].

The International Congress on Polymers in Concrete (ICPIC) is an international conference held in England and disseminated technical reports evaluating PC and its application, first in 1975. Subsequently, ICPIC congresses have been taking place every three years up to now, intending to follow the development of the application of polymer materials in the construction field [4].

1.3 APPLICATIONS OF POLYMER CONCRETE

Today, the use of PC is increasing rapidly in many industries. According to the recent research by Global Market Insights, Inc., *PC* market size is forecast to exceed USD 600 million by 2025 [7].

Due to high strength and durability, fast setting, admirable bond with the old concrete and steel substrates, high chemical stability under corrosive environment (acid, salt, etc.), and high resistance to water penetration for a long time, as well as good freeze–thaw resistance in comparison with the conventional concrete and low conductivity, PCs are an attractive choice as repair materials [8,9]. PC incorporating a very low viscosity monomer can be used to repair unavailable cases [8].

Fast curing time in maintenance applications such as highway pavements is the main advantage of PCs. Therefore, overlays are of great use for PCs [10,11]. Also, they are implemented for bridge overlay and floor coverings [12]. PC overlays are widely implemented in decoration and architecture because of the ability to create thin layers, rapid preparation, and negligible permeability. Due to their chemical resistance, PC is implemented as an anticorrosive lining of sewage canals, chemical factory floors, septic tanks, and so on [13].

High mechanical strength, relatively high modulus, fast curing, easy molding, and considerable vibration damping make the PC suitable for precast applications [14,15]. PC is implemented to fabricate numerous products such as drainages, underground boxes, building coverage, acid tanks, dangerous waste containment, and tiles [12]. Polymer concretes are widely implemented in flooring blocks, due to their excellent durability and relatively low cost [7]. PCs can be reinforced with nanomaterials, fibers, rebars, and even composite sheets [16] to improve their properties.

To use PC in any particular application, the conditions of use must be considered. For example, varying climatic conditions must be considered to construct a PC panel to have no deleterious effect. The ultraviolet radiation of the sun, elevated temperatures, and thermal cycles should have little or no impact on the matrix characteristics or its color. It should also be relatively fireproof and has good short- and long-dimensional stability with insignificant creep permanent deformation [17]. A building panel should not be affected by polluted air, and the chemical solution never absorbs dirt and should be usually self-cleaned by rain, wind, and snow. The building panels should be lightweight, and the coefficient of thermal expansion (CTE) would have to work symbiotically with the other building products to which it is attached. Trying to fit into all parameters seems an almost impossible task, and usually, most of these characteristics are attained [18].

Polymer concrete was at first implemented as a cement concrete replacement. Still, it was later implemented as an alternative to other substances such as metal alloys, for example, cast iron for machine beds [19–21]. Due to its superb vibration damping characteristic, PC is the appropriate composite for primary machine segments in high-precision machines such as CNC grinding machine tool and the milling machine [22–28]. Additionally, PC is used in frame supports for MRI, X-ray, and CT-Scanners [23]. PC can be implemented in complex concrete sleepers of the high-speed train systems because it reduces the radiation of the rolling noise [29].

PC is also used in the construction of electrical insulators since the 1970s. Hollow model insulators are manufactured by PC in literature [30–32]. Except for a finite experience with 69 kV and even with 132 kV voltage levels, PC insulators are mainly used for medium voltages. Additionally, PC was applied for bushings and housings of switchgear, surge arresters, and fuses [31]. We classified some applications of PCs which are listed in Table 1.1.

Kumar and Venkatesh [54] classified PC applications in construction industries based on resin type.

1.4 ADVANTAGES AND DISADVANTAGES OF POLYMER CONCRETES

PC has advantages of high mechanical strength like tensile, flexural, compressive, impact strengths and excellent vibration damping properties, high durability

TABLE 1.1
Polymer concrete applications

Product group	Application	Important required properties
Electrical and communication	Power and communications maintenance hole [33]	Electrical insulation, waterproof, and strength
	Cable trough for the high-speed rail line [33], nonconductive, nonmagnetic support structures for electrical equipment [34]	Electrical insulation, waterproof, and durability, bending strength
	Electrical insulator, bushings, and housings of switchgear, surge arresters, and fuses [30–32]	Resistance to tracking and erosion
	Electrowinning electrolytic cells [35]	Chemical resistant, electrical insulation, waterproof, and durability
Overlay and coatings	Overlays for bridge decks, floors, pavement [36–39]	Lightweight, high skid resistance, high abrasion resistance, rapid setting, excellent bond, rapid curing at ambient temperature, freezing–thawing stability
Repair material	Old concrete and pavement quick repair [40–42]	Strong bonding with PCC, cost-effective, impermeability to water and salts, rapid curing at ambient temperature, mechanical strength
Water supply and drainage	Pipes, maintenance holes, drainage system and their joint gadgets pre-slope trench, drainpipe, small water-flow control structures, components for the animal-feeding industry [33,34,41,43,44]	Freezing–thawing stability, chemical resistance, impermeability to water and salts, mechanical strength, precast ability
Agricultural irrigation facilities	Precast flume [33,45]	Freezing–thawing resistance, chemical resistance, impermeability to water and salts, mechanical strength
Container	High-integrity container (HIC) for disposal of nuclear waste [33,46]	Mechanical strength, chemical resistance, and biological properties
	Hazardous waste containers, effluent drains, septic tanks [41]	Mechanical strength, chemical resistance, freezing–thawing resistance, precast ability
Architectural use	Artificial marble and landscape [33], composite and insulated panels [40], window sills and copings, solar tiles [18], cultured marble for countertops, lavatories, and other sanitary ware [47]	Moderate prices, easy to color and mold with complicated patterns or designs, durability, mechanical strength, weathering resistance
Building and construction	Infrastructures, dry-mounted external wall construction joints, parking garage decks [14]	Mechanical strength, thermal stability

(*Continued*)

Product group	Application	Important required properties
Machine parts	Guides, tables, and machine bases [41,48,49]	Mechanical strength, good vibration alleviating, dynamic stiffness, thermal stability, chemical resistance
High pressure and temperature media	Use of geothermal energy [41]	Mechanical strength, chemical resistance
Hydraulic structures	Dams, dikes, reservoirs, and piers [50,51]	Mechanical strength, high abrasion-resistant surface, impermeability to water and salts, freezing–thawing resistance
Transportation	Highway surfaces and bridge decks, road barriers, tunnel walls, catwalks, railroad crossings, railroad ties [20,36,40,52]	Mechanical strength, severe weather conditions, abrasion-resistant, rapid setting, excellent bond, rapid curing, freezing–thawing stability
3D printer	Use as 3D printing material [53]	Mechanical strength, chemical stability, thermal resistance, easy to color and mold with complicated patterns

performance and specific stiffness, lightweight, chemical resistance, good corrosion stability, low permeability, easy placement, fast curing at ambient temperature and rapid setting, ability to form complex shapes, excellent adhesion to most surfaces like steel and cementitious concrete, desirable long-term freeze–thaw resistance, and considerable thermal and sound insulation characteristics.

However, there are some problems with PCs. Incorporating resin binder as an alternative to Portland cement significantly increases the cost. Therefore, PC is implemented anywhere the higher price is legitimized by outstanding characteristics, low working price, or low energy necessity. It should be noted that the cost of maintenance for PC is considerably lesser compared to OPC concrete with an applied barrier coating per year of service life [8,55,56].

Another potential difficulty of PCs is their susceptibility to elevated temperatures and thermal cycles that have a considerable influence on the strength of PC, because of the inclusion of the polymer resin and the viscoelastic characteristic of the polymer [57]. The monomers and hardeners can be volatile, combustible, toxic, and harmful to human skin, and therefore, some safety concerns appear out of implementing PC. Furthermore, some PC products are hard to manipulate with conventional tools because of their strength and density. Because of implementing a low amount of polymer binder for cost reduction, it tends to be brittle, and in some cases, fiber reinforcement should be implemented. Due to the viscoelastic characteristic of polymer, PC fails subject to a constant compression loading at stress ranges higher than 50% of the ultimate strength.

REFERENCES

[1] S.F. Nabavi, *Performance of Polymer-Concrete Composites in Service Life of Maritime Structures*, University of technology, Sydney, 2014.

[2] L. Czarnecki, M.R. Taha, R. Wang, Are polymers still driving forces in concrete technology?, in: M.M.R. Taha (Ed.), *Int. Congr. Polym. Concr.*, Springer, Cham, 2018: pp. 219–225. https://doi.org/10.1007/978-3-319-78175-4.

[3] A. Jain, Polymer concrete : Future of construction industry, *Int. J. Sci. Res.* 2 (2013) 201–202.

[4] S.M. Daghash, *New Generation Polymer Concrete Incorporating Carbon Nanotubes*, The University of New Mexico, Thesis for Master of Science, 2013.

[5] R. Bedi, R. Chandra, Reviewing some properties of polymer concrete, *Indian Concr. J.* 88 (2014) 47–68.

[6] A.O. Kaeding, A perspective on 40 years of polymers in concrete history, in: *Int. Congr. Polym. Concr. (ICPIC 2018)*, Springer International Publishing, Washington DC, 2018: pp. 321–328.

[7] Polymer concrete – global market outlook (2017–2026), 2018.

[8] R. Allahvirdizadeh, R. Rashetnia, A. Dousti, M. Shekarchi, Application of the polymer concrete in repair of concrete structures: A literature review, in: *4th Int. Conf. Concr. Repair*, Dresden, Germany, 2011: pp. 1–10. https://doi.org/10.13140/2.1.4893.7925.

[9] P. Mani, A.K. Gupta, S. Krishnamoorthy, Comparative study of epoxy and polyester resin-based polymer concretes, *Int. J. Adhes. Adhes.* 7 (1987) 157–163. https://doi.org/10.1016/0143-7496(87)90071-6.

[10] A. Douba, *Mechanical Characterization of Polymer Concrete with Nanomaterials*, The University of New Mexico, Thesis for Master of Science, 2017.

[11] H. Zhang, *An Evaluation of the Durability of Polymer Concrete Bonds to Aluminum Bridge Decks*, Virginia Polytechnic Institute and State University, 1999.

[12] A.S. Momtazi, R.K. Khoshkbijari, S.S. Mogharab, Polymers in concrete: Applications and specifications, *Eur. Online J. Nat. Soc. Sci.* 3 (2015) 62–72.

[13] M.E. Tawfik, S.B. Eskander, Polymer concrete from marble wastes and recycled poly(ethylene terephthalate), *J. Elastomers Plast.* 38 (2006) 65–79. https://doi.org/10.1177/0095244306055569.

[14] K.S. Rebeiz, Time-temperature properties of polymer concrete using recycled PET, *Cem. Concr. Compos.* 17 (1995) 119–124. https://doi.org/10.1016/0958-9465(94)00004-I.

[15] K.S. Rebeiz, Precast use of polymer concrete using unsaturated polyester resin based on recycled PET waste, *Constr. Build. Mater.* 10 (1996) 215–220. https://doi.org/https://doi.org/10.1016/0950-0618(95)00088-7.

[16] K. Yeon, Y. Kim, Y. Kim, Y. Choi, Flexural behavior of methyl methacrylate modified unsaturated polyester polymer concrete beams reinforced with glass-fiber-reinforced polymer sheets, *J. Appl. Polym. Sci.* 119 (2010) 3297–3304. https://doi.org/10.1002/app.

[17] G. Woltman, D. Tomlinson, A. Fam, Investigation of various GFRP shear connectors for insulated precast concrete sandwich wall panels. *J. Compos. Constr.* 17 (2013) 711–721. https://doi.org/10.1061/(ASCE)CC.1943-5614.0000373.

[18] R.C. Prusinski, W.S. Wahby, Polymer concrete solar tiles, in: *Int. Congr. Polym. Concr. (ICPIC)*, University of Minho, Funchal-Maderia, Portugal, 2010: pp. 405–412.

[19] J.M.L. Reis, A.J.M. Ferreira, The influence of notch depth on the fracture mechanics properties of polymer concrete, *Int. J. Fract.* 124 (2003) 33–42.

[20] D. Fowler, Current status of polymer concrete in the United States, in: *Int. Congr. Polym. Concr.*, Bologna, Italy, 1998: pp. 37–44.

[21] R. Poklemba, D. Duplakova, J. Zajac, J. Duplak, V. Simkulet, D. Goldyniak, Design and investigation of machine tool bed based on polymer concrete mixture, *Int. J. Simul. Model.* 19 (2020) 291–302. https://doi.org/10.2507/IJSIMM19-2-518.
[22] P. Xu, W. Li, Y. Yu, J. Shen, Analysis of chatter and turning errors of basalt fiber reinforced polymer concrete lathe, *J. Phys. Conf. Ser.* 1748 (2021) 1–7. https://doi.org/10.1088/1742-6596/1748/6/062058.
[23] H. Haddad, M. Al Kobaisi, Optimization of the polymer concrete used for manufacturing bases for precision tool machines, *Compos. Part B Eng.* 43 (2012) 3061–3068. https://doi.org/10.1016/j.compositesb.2012.05.003.
[24] W. Bai, J. Zhang, P. Yan, X. Wang, Study on vibration alleviating properties of glass fiber reinforced polymer concrete through orthogonal tests, *Mater. Des.* 30 (2009) 1417–1421. https://doi.org/https://doi.org/10.1016/j.matdes.2008.06.028.
[25] J. Yin, J. Zhang, T. Wang, Y. Zhang, W. Wang, Experimental investigation on air void and compressive strength optimization of resin mineral composite for precision machine tool, *Polym. Compos.* 39 (2016) 457–466. https://doi.org/10.1002/pc.
[26] S. Huicun, T. Kui, H. Yanhua, Study on the creep properties of resin concrete, *Appl. Mech. Mater.* 472 (2014) 649–653. https://doi.org/10.4028/www.scientific.net/AMM.472.649.
[27] G. Vrtanoski, V. Dukovski, Design of polimer concrete main spindle housing for CNC lathe, in: *13th Int. Sci. Conf. Achiev. Mech. Mater. Eng.*, Gliwice, Poland, 2005: pp. 696–698.
[28] J. Do, S.Æ. Dai, G. Lee, Design and manufacture of hybrid polymer concrete bed for high-speed CNC milling machine, *Int. J. Mech. Mater. Des.* 4 (2008) 113–121. https://doi.org/10.1007/s10999-007-9033-3.
[29] S. Ahn, S. Kwon, Y. Hwang, H. Koh, H. Kim, J. Park, Complex structured polymer concrete sleeper for rolling noise reduction of high-speed train system, *Compos. Struct.* 223 (2019) 110944. https://doi.org/10.1016/j.compstruct.2019.110944.
[30] L.E. Schmidt, A. Krivda, C.H. Ho, M. Portaluppi, Polymer concrete outdoor insulation – Experience from laboratory and demonstrator testing, in: *2010 Annu. Rep. Conf. Electr. Insul. Dielectr. Phenom.*, IEEE, West Lafayette, IN, 2010: pp. 1–3. https://doi.org/10.1109/CEIDP.2010.5723942.
[31] K.L. Chrzan, M. Skoczylas, Performance of polymer concrete insulators under light pollution, in: *XV Int. Symp. High Volt. Eng.*, Ljubljana, Slovenia, 2014: pp. 1–4.
[32] J.L. Fierro-Chavez, T.P. Duguid, Diagnostic techniques for the field evaluation of polymer concrete insulators, in: *Sixth Int. Conf. Dielectr. Mater. Meas. Appl.*, 1992: pp. 146–149.
[33] K.-S. Yeon, Polymer concrete as construction materials, *Int. J. Soc. Mater. Eng. Resour.* 17 (2010) 107–111.
[34] Y. Ohama, Polymer concrete, in: *Dev. Formul. Reinf. Concr.*, Woodhead Publishing Limited, Nihon University, Japan, 2008: pp. 256–269. https://doi.org/10.1533/9781845694685.256.
[35] M.E. Schlesinger, M.J. King, K.C. Sole, W.G. Davenport, Electrowinning, in: M.E. Schlesinger, M.J. King, K.C. Sole, W.G.B.T.-E.M. of C. (Fifth) E. Davenport (Eds.), *Extr. Metall. Copp.*, 5th ed., Elsevier, Oxford, 2011: pp. 349–372. https://doi.org/https://doi.org/10.1016/B978-0-08-096789-9.10017-4.
[36] M. Abokifa, M.A. Moustafa, Experimental behavior of poly methyl methacrylate polymer concrete for bridge deck bulb tee girders longitudinal field joints, *Constr. Build. Mater.* 270 (2021) 121840. https://doi.org/10.1016/j.conbuildmat.2020.121840.

[37] M.M. Sprinkel, Polymer concrete for bridge preservation, in: *Int. Congr. Polym. Concr. (ICPIC 2018)*, Springer, Cham, 2018: pp. 15–26. https://doi.org/https://doi.org/10.1007/978-3-319-78175-4_2.

[38] M. Abokifa, M.A. Moustafa, A.M. Itani, Comparative structural behavior of bridge deck panels with polymer concrete and UHPC transverse field joints, *Eng. Struct.* 247 (2021) 113195. https://doi.org/https://doi.org/10.1016/j.engstruct.2021.113195.

[39] R.J. Stevens, W.S. Guthrie, J.S. Baxter, B.A. Mazzeo, Field evaluation of polyester-polymer concrete overlays on bridge decks using nondestructive testing, *J. Mater. Civ. Eng.* 33 (2021) 4021155. https://doi.org/10.1061/(ASCE)MT.1943-5533.0003810.

[40] M.S. Stenko, Precast polymer concrete panels for use on bridges and tunnels, in: M.M.R. Taha (Ed.), *Int. Congr. Polym. Concr. (ICPIC)*, Springer, Cham, Washington, DC, 2018: pp. 353–359.

[41] J.J. Fontana, J. Bartholomew, Use of concrete polymer materials in the transportation industry, *Am. Concr. Inst.* 69 (1981) 21–44.

[42] K.C. Jung, I.T. Roh, S.H. Chang, Evaluation of mechanical properties of polymer concretes for the rapid repair of runways, *Compos. Part B Eng.* 58 (2014) 352–360. https://doi.org/10.1016/j.compositesb.2013.10.076.

[43] K.S. Rebeiz, D.W. Fowler, D.R. Paul, Polymer concrete and polymer mortar using resins based on recycled poly (ethylene terephthalate), *J. Appl. Polym. Sci.* 44 (1992) 1649–1655.

[44] L. Leonardi, T.M. Pique, T. Leizerow, H. Balzamo, C. Bernal, E. Agaliotis, Design and assessment of a lightweight polymer concrete utility manhole, *Adv. Mater. Sci. Eng.* 2019 (2019) 5234719. https://doi.org/https://doi.org/10.1155/2019/5234719.

[45] K. Seok Yeon, M. Kawakami, Y. Sang Choi, J. Yong Hwang, S. Ho Min, J.H. Yeon, Remodeling of deteriorated irrigation aqueducts using precast polymer concrete flume, *Adv. Mat. Res.* 687 (2013) 35–44. https://doi.org/10.4028/www.scientific.net/AMR.687.35.

[46] H. Chung, M.S. Lee, D.H. Ahn, H.J. Won, H.S. Kang, H.S. Lee, S.P. Lim, Y.E. Kim, B.O. Lee, K.P. Lee, B.Y. Min, J.K. Lee, W.S. Jang, W.B. Sim, J.C. Lee, M.J. Park, Y.J. Choi, H.E. Shin, H.Y. Park, C.Y. Kim, Development of polymer concrete radioactive waste management containers, Republic of Korea, 1999. http://inis.iaea.org/search/search.aspx?orig_q=RN:31035890.

[47] Y. Ohama, Recent progress in concrete-polymer composites, *Adv. Cem. Based Mater.* 5 (1997) 31–40. https://doi.org/10.1016/S1065-7355(96)00005-3.

[48] N. Kepczak, Influence of the addition of styrene-butadiene rubber on the dynamic properties of polymer concrete for machine tool applications, *Adv. Mech. Eng.* 11 (2019) 1–11. https://doi.org/10.1177/1687814019865841.

[49] M. Chod, P. Dunaj, B. Powa, S. Berczy, T. Okulik, Increasing lathe machining stability by using a composite steel – Polymer concrete frame, *CIRP J. Manuf. Sci. Technol.* 31 (2020) 1–13. https://doi.org/10.1016/j.cirpj.2020.09.009.

[50] J.M.L. Reis, A comparative assessment of polymer concrete strength after degradation cycles, in: H. da C. Mattos, M. Alves (Eds.), *Second Int. Symp. Solid Mech.*, Brazilian Society of Mechanical Sciences and Engineering, Rio de Janeiro, Brazil, 2009: pp. 437–444.

[51] D. Fowler, State of the art in concrete–polymer materials in the U.S, in: M. Maultzsch (Ed.), *Int. Congr. Polym. Concr.*, Berlin, Germany, 2004: pp. 597–603.

[52] I. Mantawy, R. Chennareddy, M. Genedy, Polymer concrete for bridge deck closure joints in accelerated bridge construction, *Infrastructures.* 4 (2019) 31. https://doi.org/10.3390/infrastructures4020031.

[53] M. Krčma, D. Škaroupka, P. Vosynek, T. Zikmund, J. Kaiser, D. Palousek, Use of polymer concrete for large-scale 3D printing, *Rapid Prototyp. J.* 27 (2021) 465–474. https://doi.org/10.1108/RPJ-12-2019-0316.

[54] G.B. Ramesh, U.G. Student, Review on performance of polymer concrete with resins and its applications, *Int. J. Pure Appl. Math.* 119 (2018) 175–184.
[55] O. Bozkurt, M. Islamoğlu, Comparison of cement-based and polymer-based concrete pipes for analysis of cost assessment, *Int. J. Polym. Sci.* 2013 (2013) 921076. https://doi.org/10.1155/2013/921076.
[56] S. Sakhakarmi, *Cost Comparison of Cement Concrete and Polymer Concrete Manholes in Sewer Systems*, University of Nevada, 2017.
[57] O. Elalaoui, E. Ghorbel, V. Mignot, M. Ben Ouezdou, Mechanical and physical properties of epoxy polymer concrete after exposure to temperatures up to 250°C, *Constr. Build. Mater.* 27 (2012) 415–424. https://doi.org/10.1016/j.conbuildmat.2011.07.027.

2 Materials of Polymer Concrete

Abstract

Polymer concrete (PCs) are the combination of resins and fillers that depending on the application and properties, a variety of aggregates, micro fillers, nanoparticles, and fibers can be used as fillers. Knowing the components, their amount, and also their effect on the properties of PC will help to achieve the final product with the expected properties. In this chapter, based on previous research, resins and fillers used in PC are introduced and classified and the effect of their variables on the properties of concrete will be discussed. Resin type and weight percentage; aggregate type, geometry, size and weight percentage; and preparation method, as material variables also will be discussed.

2.1 RESIN

Polymer resins are divided into two types thermoset and thermoplastic, according to cross-linking during processing [1]. Thermosets, are highly cross-linked polymers and are widely used in composite structures because of their excellent stability under high temperatures and a wide variety of conditions. Initial viscosity (ability to incorporation into the reinforcement) and pot-life (time needed for curing before demolding) are the two most crucial processing parameters for thermosets [2].

On the other side, thermoplastics are uncross-linked polymers. These polymers soften and flow when heated and, therefore, aren't suitable for structural applications. Melting temperature and melting viscosity are the main processing parameters of thermoplastics [2]. Thermosetting polymers are generally implemented as the binder in polymer concrete (PC) because of their remarkable thermal resistance; thermoplastic resins are also implemented to a lesser extent [3].

2.1.1 Resin Types

It is essential to pay attention to different factors like cost, appropriate mechanical and thermal characterizations, and chemical stability to choose a type of polymer. Thermosetting polymer resins generally implemented in PC are epoxy, polyester, methacrylate, and furan.

2.1.1.1 Polyester Polymer Concrete

Unsaturated polyester (UP) resin is the most widely applied polymer binder in PC due to its considerable mechanical properties, low price, and abundance [4]. Three classes of polyester resins are implemented in PC. The first type is resistant to mild corrosives and non-oxidizing mineral acids. The second one is the isophthalic type that has more stability than the previous type. The third type is based on bisphenol-A orthophthalic and has the most outstanding general stability to chemical attacks [5].

DOI: 10.1201/9781003326311-2

In most applications, UP pre-polymer formulation is used. The polyester pre-polymer and the monomer cross-links via their unsaturated groups during polymerization [6].

PC with polyester binder has good mechanical strength, remarkable chemical and freeze–thaw stability, and considerable adhesion to other materials [7]. A significant disadvantage of polyester PC is the great curing and post-curing shrinkage (up to ten times that of ordinary Portland cement (OPC) concrete). It was reported that 20 wt% polystyrene shrinkage reducing agent was mixed with 80 wt% unsaturated polyester resin to reduce PC shrinkage [8]. Another drawback of unsaturated polyester resin is its high viscosity which limits its workability. The addition of a dilutant monomer can enhance the workability of UP [9]. Polyester resins do not have good bond strength to wet and humid surfaces. They are generally flammable at 38°C, and the curing agent implemented for polyester resins disintegrates rapidly over 32°C and is at risk of fire or explosion [10]. In some works, methyl ethyl ketone peroxide (MEKP) and cobalt naphthenate have been used as an initiator (catalyst) [11–17] and promoter [18–22] of the polymerization process of polyester-based PC, respectively.

2.1.1.2 Epoxy Polymer Concrete

Epoxy is one of the most widely used thermoset resins with significant properties like good mechanical and chemical strength, remarkable stability in corrosive solutions and media, considerable resistance in outdoor environments, good bonding strength to most materials, Insignificant curing shrinkage, great versatility, remarkable creep properties and fatigue strength, and negligible moisture absorption [23]. Different curing agents are utilized as hardeners of epoxy resins: polyamines (e.g., tertiary polyamines), polyamides, and polysulfide polymers. Polyamines are the most frequently used hardener in PC fabrication and give PCs with the highest chemical stability. Polyamides epoxy PC has greater flexibility, higher thermal strength, and decreased chalking tendency in an outdoor environment, but their solubility and chemical stability are lower compared to polyamines epoxy PC. Polysulfides epoxy PC has greater flexibility than the other types [24].

Generally, epoxy PC exhibits good mechanical properties, good bonding with most substrates, low curing, and post-curing shrinkage, excellent chemical stability [10], and significant creep and fatigue strengths [23,25]. However, the cost of epoxy PC is high, and they are implemented in particular applications, which the higher price can simply be vindicated [6]. As reported in the literature, epoxy is preferred to polyester due to the superior mechanical strength as well as more excellent chemical resistance. Some characteristics of polyester PC can be improved up to the same level by reinforcement with micro fillers and silane-coupling agents [26].

2.1.1.3 Furan Polymer Concrete

Furfuryl alcohol (FA) and furfuraldehyde are the major materials implemented in the fabrication of furan polymers. FA and furfuraldehyde are thermoset bio-resins obtained from agricultural wastes. Thus furan polymer is comparatively low price in comparison to the other synthetic resins. Among different types of furan resins like homopolymer of furfuryl alcohol, a copolymer of furfuryl alcohol, and furfuraldehyde and condensation products of furfuraldehyde and acetone, the last class is more commonly used [6,27].

FA is polymerized into polyfurfuryl alcohol (PFA) by using an acid catalyst. Conversion of FA to PFA can be implemented to synthesize PFA-based concrete with significant mechanical properties with the same level of epoxy and polyester PCs and also considerable chemical stability against acid and alkali [28].

Furan PC, when properly cured, exhibits superior chemical resistance and excellent stability in high temperatures as well as intense thermal shock. Furan PC products have good resistance to most aqueous acidic or basic solutions and strong solvents such as ketones, aromatics, and chlorinated compounds [27]. However, it does not work with an alkaline surface [29]. Reinforced furan PC is used in construction applications where corrosive situations are confronted (e.g., power plants, steel mills, paper industry). Despite good properties, only a few work apply furan resins in PC [27,30,31].

2.1.1.4 Acrylic Polymer Concrete

Polymethyl methacrylate (PMMA) is obtained by polymerizing methyl methacrylate (MMA) and implemented as the most common acrylic polymer. Acrylic-based PCs have good chemical and weathering resistance and waterproofing properties, approximately slight curing shrinkage (0.01–0.1%), high freeze–thaw resistance, simplicity of viscosity amendment, considerable abrasion resistance, and relatively equal coefficient of thermal expansion to OPC concrete [32]. However, MMA-based PCs have limited use due to the low flash point (11°C) of their polymer, higher flammability, and obnoxious odor [24]. PMMA PCs are considered due to their remarkable operability and low curing temperature [33,34]. The rate of strength development in acrylic PC is significantly high at very early ages, even at subzero temperatures. The compressive strength of PMMA PC can be enhanced by adding silane [34].

Another research used an acrylic resin with benzoyl peroxide (BPO) initiator and *N,N*-dimethylaniline (DMA) accelerator as a resin of PC to evaluate deformation behavior of acrylic PC. Methacrylic acid (MAA) was implemented as an auxiliary accelerator. Increasing the MAA content increased ultimate setting shrinkage and CTE. The resin content also affected the compressive strength, ultimate strain, and elastic modulus of acrylic PC [35].

2.1.1.5 Vinyl Ester Polymer Concrete

Vinyl ester resin has a range of intermediate properties compared to epoxy and polyester resins and cost and is mainly used for particular purposes [36]. Vinyl ester exhibits high resiliency and toughness. Also, it has higher stability to hydrolysis, lesser peak of exotherms, higher curing time, and lower curing shrinkage compared to epoxy and polyester resins [13]. Also, it has good adhesion to wet or humid surfaces [10]. The vinyl ester PC exhibits lower compressive strength and CTE range than polyester and epoxy PCs. However, vinyl ester PC demonstrates deterioration because of cracks created by temperature variations and destruction of adhesion bonding between the PC and substrate and also is hard to handle [37]. Vinyl ester is modified by adding an MMA monomer for reducing viscosity and increasing workability for implementation in PC [38].

Table 2.1 represents the general properties and applications of PCs based on the resins mentioned above.

Only a few studies considered implementing polyurethane resin in PC [39–41]. As an innovative work, liquefied wood waste is used to fabricate polyurethane foam

TABLE 2.1
General properties and applications of PC products

Resin	General properties	Applications
Epoxy	Good bond adhesion to most construction materials, low curing shrinkage, excellent chemical stability, remarkable creep, and fatigue properties, low water absorption	Mortar for industrial flooring, precast PC shapes, solid resin, and Terrazzo flooring, anchor fixings, highways skid-resistant overlays, trenches, sumps, epoxy plaster for exterior walls, and coating of degraded structures
Polyester	Remarkable mechanical properties and adhesion to the other materials, desirable chemical and freeze–thaw stability, low cost, but a high setting and post-setting shrinkage	FRP bridge sections, panels of buildings, bridge deck overlays, cladding panels, floor tiles, sinks, surfaces, sewer pipes coatings, stairs, precast, and cast-in parts in the construction industry
Furan-based polymer	Considerable chemical stability (not acidic or basic solutions) good endurance in polar organic liquids like ketones, aromatic hydrocarbons, and chlorinated compounds	Brick, chemical-resistant floors and linings, high temperatures and thermal shocks, chemical-resistant pump pads, equipment pads, curbing, reaction vessels, trenches, and sumps, pits, catch basins
Poly (methyl methacrylate)	Low water absorption and therefore remarkable freeze–thaw stability; low rate of setting and post-setting shrinkage; excellent chemical stability; and outdoor permanence	Production of stair units, surfaces, sinks, facade plates, sanitary products for curbstones, structural patching material for repairing large holes in bridge decks
Vinylester	Ability to cast into shapes such as trenches and sumps, high resiliency and toughness, good resistance to hydrolysis, lower peak exotherms, less curing shrinkage	Chemical-resistant sumps, dikes, containment areas, trenches, walls and other structural support columns or bases, tank legs and piers, overlay on to extremely pitted concrete to rebuild and chemical proof in one step

resin and implemented in the manufacturing of porous PC [42]. The density of the presented concrete reinforced by wooden material as aggregates was about 250 kg/m^3. A comparative experimental study on the effect of epoxy, rigid polyurethane, and polyurethane foam resins on the mechanical strength of the PC revealed that epoxy-based PC provides more specific strength than the other two types [40]. Polyurethane resin can significantly increase the impact times and impact energy absorption of PC [41].

Sometimes a combination of more than one type of polymer is used to manufacture PC. Polyurethane resin can be combined with orthophthalate unsaturated polyester to form interpenetrating polymer networks (IPNs) in new polymer mortars [43]. MMA resin can be implemented as a diluent of unsaturated polyester PC to improve the workability and working life [44].

Many works have been done on the recycling of waste polymer and applying them into PC due to environmental and ecological considerations, lower energy consumption, and lower production cost [45]. Post-consumer polyethylene terephthalate (PET) wastes are an appropriate instance of these materials [46]. Different authors achieved the chemical transformation of recycled PET into unsaturated polyester and applied that as a PC binder [19,47–52]. As an example, oligomers were achieved by depolymerization of PET bottles and reacted with maleic anhydride and adipic acid to produce the polyester resin, and it was implemented in PC [19]. Also, the usability of recycled refuse low-density polyethylene (LDPE), high-density polyethylene (HDPE), and polypropylene (PP) was proved [53]. It was found that the developed PC using waste expanded polystyrene (EPS) solution-based binders has similar physical properties and durability to commercial unsaturated polyester PC [54].

Chung and Hong [55] fabricated elastic rubber concrete comprising waste tire solution as a way of solving the environmental problem of waste tires. The obtained strength and weathering resistance of the presented PC proved that it would be a high-strength PC with resistance to chemicals and outdoor exposure. For increasing the compliance of epoxy PC, tire waste powder and liquid-type silicone rubber were used instead of some epoxy resin for rapid repair of runways. It was shown that tire waste powder, due to the higher ductility factor has a more significant effect on the mechanical properties of the PC than the silicone rubber [56].

Thermoplastic resins differ from thermosets, in that a hardened polymer can be changed to a moldable or liquid state through the application of heat. Therefore, thermoplastic resins had not been considered for use in PC. Nevertheless, Prusinski et al. implemented two types of non-recyclable thermoplastic materials, acrylonitrile-butadiene-styrene (commonly referred to as ABS plastic) and vinyl, to fabricate industrial floor blocks [57].

2.1.2 Resin Weight Content

Regarding the fact that the use of polymer resins as binders in PC increases the cost of construction, it should be reduced as much as possible in concrete. If the amount of resin is too low, the properties and durability of concrete will be significantly reduced [58]. The amount of resin in the PC should be sufficient to cover all fillers and create proper adhesion. Hassani Niaki et al. [59] fabricated epoxy basalt PC. They illustrated that in the PC with the lower amount of binder, the aggregate-binder separation occurred, which demonstrates the lack of adequate adhesion bonding between the crushed basalt and epoxy (Figure 2.1a). By increasing the resin content, PC mortar became more compact (Figure 2.1b) and fully covered the aggregates. However, by implementing more resin, resin overload on the aggregates occurred (Figure 2.1c).

The optimum weight content of the resin is influenced by the geometric shape and dimension of the utilized aggregates. Implementing fine filler and aggregates, maybe increase the required polymer binder, up to 40% of weight content, because of the large surface area [60–63]. Generally, the polymer weight fraction (mix of resin and hardener) varies from 5 up to 15 wt% of the mixture. As an example, an investigation of the influence of polyester amount on the performance of PC determined that implementing 14–16% resin content by weight maximized flexural and compressive strength [64].

FIGURE 2.1 Surface morphology of the PCs: (a) 22.5 wt% epoxy/77.5 wt% basalt, (b) 25 wt% epoxy/75 wt% basalt, and (c) 27.5 wt% epoxy/72.5 wt% basalt PC [59]

2.1.3 Effects of Resin on the Performance of PC

The influence of the resin on the characteristics of the concrete does not only include resin parameters such as physical and chemical properties, content, curing, and so on. It is primarily affected by the generated interfacial bonding during the synthesis of the PC. In Table 2.2, the influence of resin variables on the characteristics of PCs is presented.

2.2 AGGREGATE

Aggregates are inert organic materials or are manufactured from industrial by-products that are distributed throughout the resin in PC. The aggregates must be clean (dust-free) and dry to enhance the polymer-aggregate bond strength to achieve maximum performance [3]. Aggregate type, size, shape or geometry, and weight content are effective aggregate variables that influence PC properties. These variables are discussed in this section.

2.2.1 Aggregate Types

Different kinds of aggregates like siliceous sand [65,70–72], river sand [73,74], Ottawa sand [75], alluvial sand [76], quartz [77,78], limestone [79–81], Pea gravel [82,83], granite [78,84,85], calcium carbonate ($CaCO_3$) [86], crushed basalt [59,65,87,88], spodumene [87], marble [89], ceramsite lightweight aggregates [90,91], lightweight expanded clay aggregate (LECA) [92], pumic [93], and chalk [87,94] were used in PC structure. Each of them has specific characteristics due to their chemical

TABLE 2.2
Effects of resin variables on PC properties

References	Resin type	Aggregate/ micro filler	Resin variables	Studied properties	Results
[65]	Epoxy	Silica sand	Resin wt%	Compressive, splitting tensile and flexural strengths	1. Maximum compressive, splitting tensile and flexural strengths were achieved using 24, 26, and 24 wt% epoxy resin, respectively
[59]	Epoxy	Crushed basalt	Resin wt%	Compressive, flexural, and splitting tensile strengths	1. Maximum strengths were obtained by 25 wt% resin
[64]	Epoxy, polyester	Coarse and fine aggregate	Resin wt%	Compressive and flexural strengths	1. Highest compressive strength was obtained by 12 wt% resin for all types of resins 2. Incorporating 13 wt% resin led to the maximum modulus of rupture (about three times of OPC concrete)
[32]	MMA-PMMA	Crushed silica stone/$CaCO_3$	MAA wt%, curing temperature, and curing age	Compressive, flexural, and splitting tensile strengths, elastic modulus, and Poisson's ratio	1. MAA can be used at a low temperature and has a short curing age of 12 hours 2. Increasing MAA wt% increased the modulus of elasticity and Poisson's ratio 3. Acrylic PC can be fabricated at low temperatures of −20°C
[66]	Polyester	Clean sand and foundry sand/$CaCO_3$	Resin wt%	Flexural strength	1. Best flexural properties were achieved for 20 wt% resin
[67]	Epoxy	Gravel and sand	Resin wt%, curing conditions	Compressive and flexural strengths	1. Incorporating 13 wt% resin led to maximum compressive and flexural strengths 2. The maximum strengths were obtained after curing for three days

(Continued)

Effects of resin variables on PC properties

References	Resin type	Aggregate/ micro filler	Resin variables	Studied properties	Results
[27]	Furan	Silica sand/quartz	Resin wt%, different catalyst system, promoter content	Density, water absorption	1. A composition of p-toluene sulfonyl chloride and sulfamic acid was selected to have a synergistic influence on furan-based PC 2. An optimum properties were obtained by 0.5 wt% silanes 3. The maximum density was obtained by 9.375 wt% resin 4. PC incorporating 9.375 wt% furan exhibited higher water absorption even though it is denser than PC including 7.5 wt% furan
[68]	Polyester	Crushed stone and fine sand	Silane treatment and resin wt%	Compressive strength and compressive fatigue	1. Polyester amount does not have a considerable influence on compressive strength 2. For a frequency range of 200–400 Hz, the temperature rose 3. Incorporation of 1% silane agent enhanced the load for resisting 2 million cycles from 59% to 64% of ultimate strength
[26]	Epoxy, polyester	Crushed quartzite, siliceous sand/$CaCO_3$	Resin type and silane treatment	Compressive, flexural, and splitting tensile strengths	1. Epoxy PC exhibited higher strengths than the polyester PC 2. Addition of a silane-coupling agent increases the compressive strength of epoxy and polyester PC by 36% and 30%, respectively

[28]	Polyester		Curing systems (MMA and maleic anhydride)	Compressive strength	1. Polyester PC had higher compressive strength in both dry and wet states than polyester PC with isophthalic acid, with these curing agents
[25]	Epoxy	Fire retardant filler, hollow microsphere, and fly ash	Resin and hardener content	Exothermic reaction	1. The higher resin content produced more heat in the mixture 2. When the filler amount was more than 40%, heat production is negligible and could be absorbed by fillers
[30]	Furan	Sand	Resin content, promoter content	Compressive strength and load-bearing	1. Increasing resin amount increased load-bearing and compressive strength of PC 2. Increasing promoter content, enhanced compressive strength, and load-bearing of PC 3. PFA concrete shows higher compressive strength and load-bearing capacity in comparison to cement concrete
[44]	Polyester and MMA	Fine aggregate/ filler (unknown)	Polyester: MMA ratio	Compressive strength, flow test, Working life, shrinkage	1. An increase in MMA amount improved the workability and working life and relatively decreased the setting shrinkage 2. Incorporating more MMA decreased the compressive strength and also modulus of elasticity, but the decrease was much smaller than the ones synthesized by a styrene monomer

(Continued)

References	Resin type	Aggregate/ micro filler	Resin variables	Studied properties	Results
					3. Implementing MMA as the diluent of high viscosity polyester has more advantages compared to the styrene monomer
[69]	Vinyl ester	Standard sand and natural gravel/waste mineral powder of asphalt and quartz powder	Resin content	Consistence and temperature variations during setting of polymer mix, an optimal combination of vinyl ester, and aggregates	1. When the waste powder content increased, the subsequent steps of setting became longer, and the temperature of micro-mortar became lower. Also, the fluid consistency decreased, and the denser mix was obtained 2. The optimal compositions of polymer to micro filler mass ratio and the waste powder to micro filler mass ratio were achieved 0.571 and 0.146, respectively

compositions. However, authors sometimes utilize locally available aggregates like waterway sand and squashed stone [95] and rock aggregate [84].

Currently, the reuse of solid waste is a vital environmental concern, and various solid wastes are implemented instead of aggregates in the manufacture of recycled PC [96]. Solid waste aggregates that implemented in PC are destruction materials [97], crushed cement/concrete, ceramic tiles and building blocks [98,99], crushed brick [100], blocks waste of concrete sleepers [80], crushed PC, rock dust, and coal mine waste [101], waste marble powder [102], glass-fiber-reinforced polymer (GFRP) waste [103], residual foundry sands [104,105], glass waste [106,107], scrap tire rubbers (crumb rubber) [88,108], waste container glass (WCG) [44], waste auto glass (WAG) [44], waste Tetra Pak particles from beverage containers [109,110], wood wastes [42], saw dust from the wood industry [111], red mud (alkaline waste from bauxite processing) [60], machining chips (steel, aluminum, and titanium chips) [112], silica fume and fly ash from industry development and electrical production wastes [113], electronic plastic waste [114], cork granulate (by-product within the cork industry) [115], metallurgical wastes (Ladle Slag and Alumina Filler) [116,117], reclaimed asphalt pavement [100], and rapid-cooled slag produced in the steel manufacturing process [8].

2.2.2 Aggregate Size

From the previous studies, the appropriate aggregate size (fine or coarse) and aggregate size distribution in PC are according to a theoretical foundation desirable for OPC concretes. PC aggregates grading is non-standardized until now. Aggregates are classified into two size groups, fine aggregates (less than 5 mm) and coarse aggregates (larger than 5 mm). An appropriate aggregates size is a parameter that is affected by the dimension of the product and required thickness and density [10].

Selecting an appropriate grad of aggregates for PC would assure a minimum void volume fraction minimizes the amount of resin needed for bonding fillers and aggregates and, therefore, reduces production cost [3]. Several optimized mix proportions based on various optimization criteria such as minimum void content, Fuller's curve, and maximum bulk density (MBD) were discussed for achieving minimum void content and MBD [118,119]. The effect of aggregate ratio on the formation of voids and a procedure of optimizing the mixture ratio to obtain the least void content was discussed in Ref. [120]. The obtained results exhibited that better mechanical properties are obtained due to the utilization of gap-graded aggregates.

Hassani Niaki et al. [59] experimentally studied the influence of four different aggregate sizes on the strength of the PC. Increasing the aggregate size increased the compressive strength, decreasing the flexural and splitting tensile strengths. Evaluation of the aggregate dimension as a design parameter of PC represented that coarsest aggregate yielded optimum compressive, bending, and interfacial shear strengths [71]. Research [121] on simultaneous utilization of an aggregate with different particle sizes was performed. Results proved that a combination of three silica powder gradation levels (0.05–0.06, 0.6, and 1.1 mm) achieved the highest mechanical properties. It should be noted that implementing finer aggregate due to higher contact surfaces increased polymer resin usage in the mixture.

2.2.3 Aggregate Shape

Figure 2.2 depicts SEM micrographs of commonly used aggregates in PCs. It can be seen that the aggregates have irregular geometry. Aggregates with non-uniform shape and large surface area increase the bonding with the polymer binder and enhance mechanical properties [122].

Numerical analysis of the effects of aggregate shape and distribution on the strength and fracture properties of polyurethane PC demonstrated that they have considerable effects on the microcracks formation and growth and also on the failure strength [164]. The results show that aggregates with more sharp angles induce more serious stress concentration, and therefore, microcracks are easily formed in those areas, which would decrease the strength of the PC. Besides, aggregate with smooth vertices prevents microcracks growth inside the concrete.

FIGURE 2.2 SEM micrographs of different aggregates: (a) quartz [123], (b) spodumene [124], (c) basalt [65], (d) chalk [125], (e) granite [126], (f) marble [127], (g) river sand [128], (h) Ottawa sand [129], and (i) silica sand [65]

2.2.4 Aggregates Weight Content

The weight percentage of aggregates has always been one of the crucial topics in research. Because the percentage of aggregates will play a very decisive role in the properties of concrete and its cost. Generally, due to the low price of aggregates, using a higher weight percentage will reduce the cost of concrete. However, some properties of PC may be reduced as a result. Therefore, finding the optimal value for the weight percentage of aggregates is always considered.

The weight ratio of aggregate/resin varies from 1:1 to 15:1, based on the aggregate range and monomer formulation [130]. The optimum aggregate/resin ratios of crushed basalt (with 1–5 mm aggregate size)/epoxy PC and silica sand (with 50–200 μm particle size)/epoxy PC were similarly attained 1:3 [65].

2.2.5 Effects of Aggregates on Polymer Concrete Properties

In addition to mentioned parameters like type, size, shape, and weight fraction, some other parameters may have a significant effect on PC properties. Aggregates must be dry or a little bit damp. Any aggregate contamination such as dust and moisture decreases polymer-aggregate interfacial bonding [10]. In addition to aggregates strength, appropriate polymer-aggregate interfacial bonding determines the PC strength. For example, from the SEM micrographs of the fracture surface of epoxy-silica sand PC, there is a desirable binder/aggregate bonding, and the strength of silica is higher than the strength of the epoxy. Therefore, the PC is broken down from the polymeric regions (see Figure 2.3).

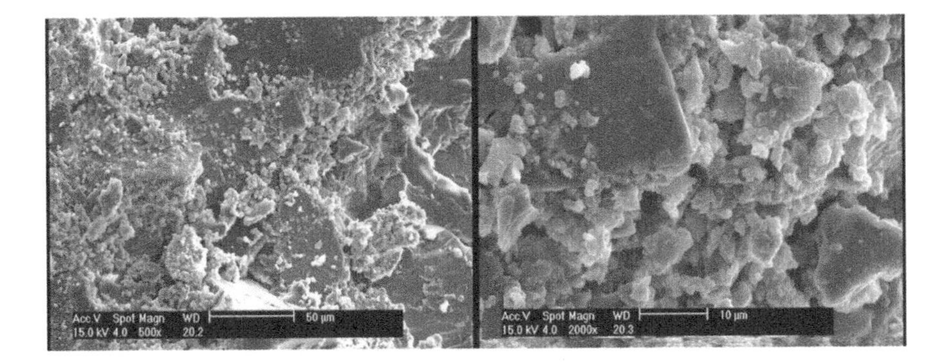

FIGURE 2.3 SEM micrographs of the fracture surface of the epoxy-silica sand PC [65]

The inclusion of an aggregate may have significant effects on different properties of PC. As an example, the addition of crumb rubber from the waste tire to an epoxy/basalt PC enhanced the compressive and the splitting tensile strengths by 9% and 8%, respectively, enhanced the PC/substrate adhesion bonding and decreased the thermal conductivity of the PC [88]. Implementing basalt and limestone waste aggregates manufactured in the recycling process of concrete sleepers, in the polyester PC, presented that recycled basaltic aggregate improved the mechanical properties of PC more than recycled limestone aggregate [80].

PCs can be utilized in more than one type of aggregate. For example, 24.7 wt% quartz fine sand, 49.3 wt% quartz gravel, 6.7 wt% quartz powder, and 7 wt% chalk in 12.3 wt% orthopthalic polyester resin, achieved PC with higher values of mechanical performance [122]. An experimental study showed that a PC made of epoxy, fly ash, silica sand (fine aggregate), and crushed basalt (coarse aggregate) has better mechanical strength than each of the binary composites (i.e., epoxy-fly ash, epoxy-silica sand, and epoxy-crushed basalt). Implementing the aggregates and fillers with various dimensions and geometry decreased the cavities stress concentration region, caused an excellent package of the mixture, and increased the strength of the PC [65]. A similar conclusion was reported by Mohamed et al. [61]. Combining two sizes of sand aggregate enhanced the flexural strength of PC and achieved denser concrete that attributed to the decrease of voids between the aggregates. It can be noted that the aggregate mixture with the highest bulk density and minimum void volume results in the maximum strength [131]. For achieving the maximum bulk density, different content of aggregates should be mixed [119] according to the ASTM C29 standard.

SEM micrographs of the fractured surface of PC containing one or more types of aggregates proved the presence of cavities or microvoids (as seen in Figure 2.4). Microvoids lead to stress concentration area that reduces concrete strength. Although it is possible to limit the formation of microvoids, their presence due to the unregulated geometry of the aggregates and bubbles creation during the mixing process in polymer resins is unavoidable. Some authors used a combination of different aggregate sizes as fine and coarse aggregates to fill the microvoids that will increase the

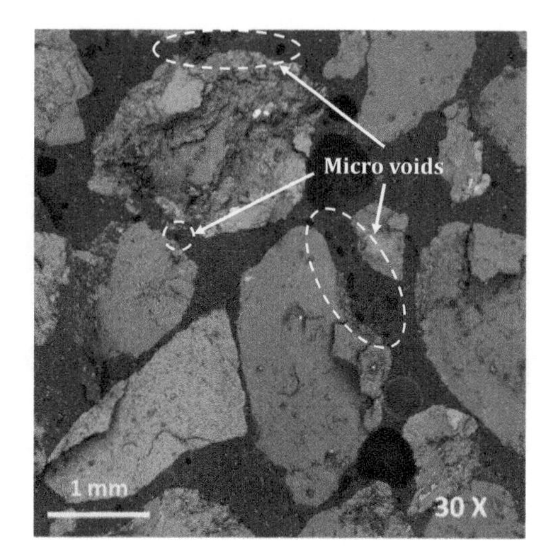

FIGURE 2.4 SEM photograph from the fracture surface of epoxy/basalt PC [59]; micro-voids are visible

mechanical strength and chemical resistance of PC [61,65,71,122]. However, micro filler usage is the best approach. In Table 2.3, we review the effects of aggregates and their variables on PC properties.

2.3 MICRO FILLER

The micro filler is a fine powder with a particle dimension smaller than filler or aggregate. Micro fillers are introduced to the PC mixture to fill the microvoids (space between the fine and coarse aggregates). Therefore, applying micro filler in PC reduces the total void volume and average pore size and thus, improves the physical and mechanical characteristics of PC. Also, resin consumption decreases, and it reduces the fabrication cost of PC. As micro fillers fill the pores between the aggregates, the polymer can concentrate on the aggregate interfaces rather than in the voids between the aggregates [26].

The micro filler provides a larger surface area than aggregate at the equivalent weight and, therefore, provides an enormous prospect for micro filler-polymer interaction that increases the mechanical strength of PC. It was found that using micro filler attains denser concrete that undergoes delayed diffusion of aggressive materials [133]. This section reviews the effects of micro filler variables such as type, size, weight fraction on PC properties.

2.3.1 Micro Filler Types

Different kinds of micro filler have been used in PC up to now. Using inorganic micro fillers, by-products, and waste materials into PC leads to cost reduction and strength improvement [134]. Fly ash [13], silica fume [113], and $CaCO_3$ [90,135–137]

TABLE 2.3
Effects of aggregate variables on PC properties

References	Resin	Micro filler	Aggregate variables	Studied properties	Results
[59]	Epoxy	–	Crushed basalt, content: 70, 72.5, 75, 77.5, and 80 wt%), size: less than 0.6, 0.6–1.2, 1.2–3, 3–5 mm	Compressive, flexural, and splitting tensile strengths	1. Maximum compressive, flexural, and splitting tensile strength achieved by 75 wt% basalt 2. Compressive strength increased by increasing the aggregate size. Flexural and splitting tensile strengths had contradictory behavior compared to the compressive strength
[30]	Furan	Quartz	Sand, size: 180–210, 210–355, 355–420 μm	Compressive strength, load-bearing	Increasing aggregate size decreased compressive strength
[65]	Epoxy	Fly ash	Basalt, content: 70–80 wt%; Silica sand, content: 72–84 wt%	Compressive, flexural, and splitting tensile strengths	1. The highest compressive and flexural strength was obtained by 76 wt% silica sand, and the highest splitting tensile strength was obtained by 74% silica sand in epoxy-silica sand PC 2. An optimum quaternary PC was obtained by 25 wt% epoxy, 5 wt% fly ash, 52.5 wt% silicious sand, 175.5 wt% crushed basalt aggregate 3. The quaternary PC has better mechanical properties than each binary composite

(Continued)

References	Resin	Micro filler	Aggregate variables	Studied properties	Results
[132]	A 3:2 volumetric ratio of unsaturated polyester to MMA, 0.8% cobalt octoate (promoter), 0.2% DMA (accelerator) and 2% (v/v) MEKP (initiator)	_	Basalt; spodumene; fly ash; river gravel; sand and chalk; content and combination	Flexural strength, coefficient of thermal expansion (CTE)	1. The best combination (minimum CTE and remarkable flexural strength) is basalt, sand, and fly ash (87% filler and 13% resin) 2. Reducing the resin reduces the negative effect of aggregates on CTE. and the resulting flexural strength was admissible 3. Replacing any aggregate of the optimum PC with another (gravel, sand, and chalk) causes a further decline in flexural strength and an increase in CTE due to the interfacial adhesion bonding behavior of particles
[8]	Unsaturated polyester, polystyrene (anti-shrinkage), MEKP (catalyst), and cobalt octoate, accelerator, respectively	Fly ash	Fine aggregate: River sand, content: 0, 5, 10, 15, 20 wt%, size: 0.1–1.2 mm; RCSS, content: 0, 5, 10, 15, 20 wt% size: 0.1–1.2 mm; Coarse aggregate: crushed stone, content: 51, 51.5, 52, 52.5 wt%, size: 5–8 mm	Compressive and flexural strengths, hot water resistance, absorption, pore characteristics	1. Increasing the resin and the replacement ratio of RCSS decreased the absorption rate (due to more densified PC) and increased compressive and flexural strengths. More than 9.0 wt% resin and 50 wt% RCSS led to a sudden reduction in compressive strength 2. Compressive and flexural strengths were reduced significantly after the hot water resistance test. PC incorporating RCSS could be damaged under the elevated temperature condition

References	Resin	Micro filler	Aggregate variables	Studied properties	Results
					3. Increasing the RCSS ratio, enhanced the bulk density and pore volume, porosity, and average pore diameter decreased
[44]	Epoxy	Fly ash	Waste container glass (WCG), size: 0–1.4 mm; waste auto glass (WAG), size: 0–1.6 mm; waste foundry sand (WFS), size: 0–1 mm; Waste slag (WS), size: 0–1.6 mm; filter fly ash polluted by flue gas denitrification (FAD) size: 0–0.36 mm; content: 50–85 wt%	Compressive and Flexural Strength, Pull-off bond strength	1. The most suitable substitution for the silica sand was WFS and WCG 2. The most optimal fraction of these secondary aggregates is 75 wt% when the maximum compressive strength was 120 MPa and the same flexural strength was 35 MPa

are the most popular micro fillers in PC. Palm oil fuel ash (POFA) [138], rapid-cooled steel slag (RCSS) [8], milled stone (quartzite, granite, and andesite mills) [139], grinding quartz powder [140], and waste glass fiber (GGF) [141] were also used as micro filler. In some research, micron-size aggregates were used as micro fillers with another aggregate in PC [142,143]. Scanning electron microscopy images of common functional micro fillers are illustrated in Figure 2.5. Some functional micro fillers are described below.

2.3.1.1 Fly Ash

Fly ash, a powdery pozzolan substance, is a by-product of power plants and is the most usable and favorite micro filler in PC due to its excellent performance.

FIGURE 2.5 SEM micrographs of common micro fillers: (a) fly ash [144], (b) silica fume [145], (c) $CaCO_3$ [146], (d) rapid-cooled steel slag [147], and (e) palm oil fuel ash [148]

The use of the fly ash micro filler in PC gives superior packing of the mortar, improves mechanical strength, chemical resistance, and surface finish of the product, reduces intermolecular friction, enhances workability, facilitates mixing, and also decreases water absorption of PC [13,65,132]. Garbacz and Sokołowska [139] investigated and analyzed the influence of fly ash on mortar properties, workability, and hardened PC properties (mechanical and physical characteristics, chemical and thermal stability).

From the SEM image of Figure 2.5a, fly ash micro particles geometry is the spherical shape, and as reported, they have a "ball-bearing" effect in the concrete mixture. Also, it can be seen from the micrograph that some of the particles be tightly attached, forming particle agglomeration, which leads to non-homogeneous dispersion of microparticles into polymer binder of PC. Fly ash microspheres have considerable particle size variation but similar surface topography. SEM micrograph also confirms the relatively smooth surface of fly ash microspheres [144,149].

2.3.1.2 Silica Fume

Silica fume or micro silica, a by-product of the electric arc furnaces, is a mineral admixture of amorphous silicon dioxide with a particle size of 100–150 times finer than a grain of cement [150,151]. Silica fume improves the strength of the PC [60,113]. As illustrated in Figure 2.5b, silica fume particles comprise tiny microspheres (average diameter of about 150 nm) with relatively smooth surfaces; however, they are smaller particles than fly ash [145].

2.3.1.3 Calcium Carbonate

Calcium carbonate ($CaCO_3$) is the most commonly used micro filler in the plastics industry. The inclusion of $CaCO_3$ enhanced the mechanical properties and improved the workability of the mixture [131]. The addition of $CaCO_3$ as a micro filler improves the compressive and flexural strengths of the polyester PC. However, splitting tensile strength shows little change [26]. SEM images of $CaCO_3$ micro particles demonstrate roughness of the surface of the particles (see Figure 2.5c) [146].

2.3.1.4 Rapid-Cooled Steel Slag

Implementing steel slag in concrete construction can alleviate the need for their disposal and decreases the use of natural aggregates and micro fillers [152]. Due to its high strength, durability, and chemistry, steel slag is a valuable material in the construction industry.

RCSS particles with spherical shapes and various size distributions (see Figure 2.5d) can provide PC with high density. As presented in Figure 2.5d, the particle size of RCSS varies from 0.1 to 1.2 mm, and the surfaces of the RCSS are highly smooth and spherical. Because of providing a ball-bearing effect due to spherical shape, it increases workability, strength, and endurance as well as the cost efficiency of PC mortar [8,153].

It is observed that by replacing the aggregate with RCSS filler, the total pore volume existing in PC and the porosity decreased due to the denser tissues of RCSS than fine aggregates. Therefore, when the amount of replaced RCSS increased, the bulk density of the PC increased [154].

2.3.1.5 Palm Oil Fuel Ash

POFA, a waste by-product of palm oil mill boilers [155], gives superior filling ability, reduces drying shrinkage, increases compressive strength, the flowability of mortar, and workability of PC [138]. SEM study of POFA micro filler proves the irregular geometry and porous structure of the particles [156].

Attention to the economic and environmental issues has been led to the recycling of materials in recent years [157]. Tire waste micro filler is obtained by the fragment of the tire, removing the reinforcement insertion, making balls from rubber, and grinding them in micron size [158]. PC with tire waste was introduced as a material with good resistance to aggressive chemical agents and frost-thaw cycles [159].

Also, the authors considered the applicability of waste limestone and quartz micro powder as micro fillers of PC with the vinylester binder. It is concluded that quartz micro powder may be replaced by waste limestone powder in the large share [69].

2.3.2 Micro Filler Size

Implementing micro filler with a smaller size leads to uniform and homogenous distribution of microparticles in a polymer matrix, enhances the interface between the polymeric and the solid micro fillers, and increases the degree of cross-linking of the binder. Thus, smaller micro filler provides a better improvement in properties [134].

Raja et al. [134] discussed the influence of micro filler dimensions on the mechanical performance of PC, using 10 wt% fly ash with various particle sizes from 300 nm up to 50 μm in epoxy resin. From the obtained results, size reduction of fly ash particles increased the hardness and impact strength of PC, and the best result was achieved for 300 nm size fly ash.

2.3.3 Micro Filler Weight Content

Although the use of micro filler in PC improves its properties, exceeding the limit content values can lead to severe deterioration of properties as reported in the literature [65,139,160]. The optimal amount of micro filler should be comparable to the basic aggregate of PC or be the same in terms of petrography. The analysis of the relationship between the amount of micro filler and resin in PC estimated the appropriate ratio of resin to micro filler in the range of 0.4–0.6 [161].

An empirical investigation on the performance of epoxy-fly ash composite shows that if the fly ash/epoxy weight ratio exceeds 1:5, the compressive, flexural, and tensile strengths of the PC will decrease sharply. Applying more amount of micro filler causes the constitution of more complex aggregation of fly ash microparticles because of the intrinsic van der Waals attractions among micron-sized filler particles. The agglomeration of filler particles leads to the generation of the stress concentration areas, which are failure start points. Therefore, applying much more micro filler reduces the strength of PC [65]. From the SEM micrographs of Figure 2.6, the fly ash particles do not have excellent adhesion to the matrix, due to the smooth surface and completely spherical shape of the particles. However, due to the micron size, the contact surface of the particles with a matrix is very large, which leads to the formation of epoxy-fly ash interfacial bonding and increases concrete strength. It

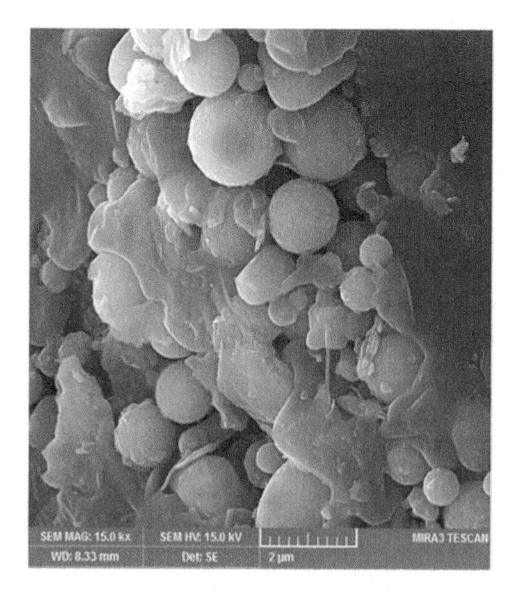

FIGURE 2.6 SEM study of the fracture surface of the 84 wt% epoxy/16% fly ash composite [65]

can be seen that the failure occurred at the matrix-particles interface and somewhere caused the particle to be separated from the matrix. But the broken particle does not appear in the image. The mechanical properties of the PC will drop if the fly ash content is so high up that there is not enough resin to encompass particles and form interfacial bonding.

Applying fly ash in PC about 12–15% of composite mass improves mechanical properties and chemical stability. However, using fly ash more than 15% decreases compressive and flexural strengths and elasticity modulus. Also, more than 20% micro filler reduces tensile strength and workability [139].

Jo et al. investigated the effect of different filler weight content from 0% to 30% in constant recycled-PET resin content. It can be seen that the implementation of $CaCO_3$ micro filler had more of an influence than its amount [135]. A comparison of epoxy and polyester PCs exhibited that the optimum $CaCO_3$-aggregates ratio is 6:94, and the $CaCO_3$-resin proportion is about 5.2:12 [26]. Implementing 7.2 up to 10.4% weight fraction of silica fume as micro filler in PC resulted in maximum compressive strength, and for the fly ash weight fraction of 12.8%, a larger value of the compressive strength was attained [113]. In another research, the silica fume weight fraction varied between 6.4% and 9.6% in PC [162].

2.3.4 Effects of Micro Filler on Polymer Concrete Properties

The present section describes the influence of micro filler variables on PC performance. A comparative experimental study on durability of a vinyl ester PC, including fluidized fly ash after 18 months and the same PC after 7 years, proved that there was no decrease in strength, but also a significant improvement of strength (probably, caused by fly ashes chemical composition) was noted, compared to the respective samples tested after 14 days [163].

Mechanical characteristics of trinary epoxy PC with river gravel as an aggregate improved by using different kinds of filler (silica fume and fly ash). By increasing the filler weight content, the compressive and splitting tensile properties were enhanced, while the flexural property slowly decreased. Fly ash particles enhanced the mechanical strength more than silica fume [113].

A comparative study of $CaCO_3$ and fly ash as a filler of polyester PC indicated that the $CaCO_3$ microparticles due to larger surface area and the superior polymer-aggregates interfacial bonding provide better creep strain and specific creep. Also, it was founded that both filler types decreased the creep strain of PC due to the significant influence of filler on restricting the deformation of PC [135].

The incorporation of fly ash and RCSS as micro filler to PC improved workability and compressive strength, as well as flexural strength of the PC. By applying more amount of polymer binder and RCSS, the PC was further densified, and thus, the absorption rate of PC tended to decrease. Also, a considerable reduction of about 21.3% in polymer binder amount in PC was reported, which caused cost efficiency by 18.5% when compared to the conventional product. However, applying RCSS caused a decrease in performance at high temperatures [8].

The inclusion of fly ash on the PC with different binders (polyester, vinyl ester, and epoxy) and sand as aggregate caused a decrease in the resin weight content and improved compressive strength and elastic modulus. However, splitting tensile and flexural strengths and ductility were reduced by incorporating more fly ash wt% [13]. Also, compared to quartz flour micro filler, fly ash had further positive effects on the mechanical properties of epoxy PC, due to the greater fineness [164].

Mirza et al. [138] considered implementing agricultural waste as a micro filler in PC. The addition of grounded POFA with dimensions smaller than 45 μm in polyester PC increased the compressive strength and the flowability of mortar. Maximum compressive strength was achieved with 14% POFA content. The results proved that POFA with physical surface modification could be implemented as a green micro filler in PC.

Diaconescu et al. [158] prepared and used a rubber waste micro filler with the average size of 68.9 μm in epoxy PC for determining the effect of micro filler and resin wt% on the mechanical properties. From the experimental study, the highest compressive and flexural strengths were obtained with 23 wt% epoxy and 30 wt% tire powder. The rest of the composite weight is composed of 30 wt% aggregates with the dimension of 0–4 mm (Sort I) and 30 wt% aggregates with the dimension of 4–8 mm (Sort II). The maximum split tensile strength was obtained with 20 wt% epoxy and 17 wt% tire powder, 33 wt% (Sort I), and 30 wt% (Sort II) [158].

A mix of quartz powder and waste perlite powder (the waste of expanded perlite production) was implemented as a micro filler in polyester PC. Perlite powder did not improve the mechanical and chemical resistance of PC [165]. The effects of micro fillers and their variables on PC properties are listed in Table 2.4 as reported in the literature.

2.4 FIBER

Due to minimizing the resin weight content, PCs exhibit some brittle properties. Features like toughness and post-peak stress-strain behavior are considered to assess the usefulness of the PC exposed to the impact, and fatigue loads [122]. A large number of studies evaluated the effect of reinforcement of PCs by incorporating synthesis or natural fibers. Fiber-reinforced polymer concrete (FRPC) is obtained by a random distribution of chopped strands fibers in PC. Different kinds of fibers and their effective performance indexes are evaluated in this section.

2.4.1 Fiber Types

Various synthesis or natural fibers like steel, glass, carbon, polyester, cellulose, fabrics, coconut, sugar cane bagasse, banana, and basalt fiber have been implemented in PC. Renewability, low cost, abundance, non-abrasive characteristics, and less concern with health and safety are some benefits of natural fibers [72].

2.4.1.1 Steel Fiber

Steel fibers can improve the post-peak load-carrying capacity of PC after initial cracking due to high strength and elastic modulus. Steel fiber-concrete bonding can be strengthened by mechanical anchorage and surface roughness of fibers. Incorporating steel fiber in PC increases the ductility, which leads to better post-peak

TABLE 2.4
Effects of micro filler variables on PC properties

References	Resin	Aggregate	Micro filler variables	Studied properties	Results
[65]	Epoxy	Silica sand, crushed basalt	Fly ash content	Compressive, flexural, and splitting tensile strengths	1. The mechanical strength is enhanced by incorporating fly ash as a micro filler 2. Maximum compressive strength was achieved by 18 wt% fly ash and maximum flexural and splitting tensile strengths at 16 w% fly ash in epoxy-fly ash composite
[82]	Polyester	Pea gravel as coarse aggregate and sand as fine aggregate	Fly ash content	Compressive strength	1. Inclusion 15 wt% fly ash instead of sand increased compressive strength by 30% 2. The relatively high content of fly ash would make the mix too sticky and unworkable (because of the high contact surface area)
[166]	Epoxy	Recycled glass sand	Metakaolin (MK) and fly ash	Compressive and flexural strengths, chemical stability, water absorption	1. MK and fly ash exhibited a significant effect on the compressive strength and flexural strength as well as the modulus of elasticity 2. The reinforced PC with MK and fly ash exhibited remarkable chemical stability for 20% Na_2CO_3, 10% NaOH, tap water, and seawater 3. The PC had low apparent porosity and low water absorption

(Continued)

References	Resin	Aggregate	Micro filler variables	Studied properties	Results
[113]	Epoxy	River gravel	Silica fume and fly ash content	Compressive, flexural, and split tensile strengths	1. Increasing the filler amount improved the compressive and split tensile strengths. However, the flexural strength was reduced with a low slope 2. Inclusion of silica fume and fly ash improved the mechanical strengths of PC 3. Inclusion of more epoxy resin, enhanced the strengths of fly ash-reinforced PC more than silica fume-reinforced PC
[135]	Polyester	Quartzite and natural aggregate	Fly ash and $CaCO_3$ filler type and content	Creep behavior	1. Introduction of filler decrease creep strain of PC 2. $CaCO_3$ demonstrates a better effect on the creep strain and specific creep than the fly ash
[138]	Polyester	Unground POFA (UPOFA) and silica sand	Palm oil fuel ash (POFA) and calcium carbonate content	Flowability, compressive strength	1. Ground POFA increased the flowability of mortar and enhanced filling ability in PC and outcomes superior compressive strength 2. Physical surface-modified POFA can be effectively used as a green micro filler in PC
[158]	Epoxy	Aggregate	Tire powder content	Compressive, flexural, and splitting tensile strengths	1. Addition of tire powder improved splitting tensile strengths 2. Increasing tire powder content decreased mechanical strength

behavior [167]. Also, an increment in compressive and flexural strengths of steel fiber-reinforced PC was reported [167,168].

2.4.1.2 Basalt Fiber

Basalt fiber is an inorganic fiber with a typical filament diameter between 10 and 20 μm. It is achieved from basalt-based igneous volcanic rocks. Basalt fiber, due to high mechanical properties and thermal resistance, excellent chemical resistance, wide operating temperature range (−200°C to 600°C), dielectric properties, low price, non-combustible and non-toxic specification, and environmental friendliness, is the most exciting fiber among the natural fibers that implemented for PC reinforcement [169,170].

The main feature of basalt fiber is its significant adaptability and bonding with the polymer matrix like epoxy, polyester, and vinyl ester. Thus, they have been implemented for the reinforcement of polymer composites like PC. Just like glass fiber, basalt fiber has a large number of silanol groups on its surface. Therefore, a good interfacial adhesion bonding between basalt fibers and hydroxyl groups of epoxy is obtained after curing [171].

As reported in the literature, basalt fiber is close to or even surpasses glass or carbon fibers in mechanical properties and physical characteristics [172,173]. Incorporating basalt fiber in concretes enhances the mechanical strength, thermal resistance as well as vibration damping [174–177].

2.4.1.3 Carbon Fiber

Carbon fiber is an extremely thin fiber that is 5–10 μm in diameter and is composed of at least 92 wt% carbon atoms. Carbon fiber is increasingly being considered because of its several advantages, including high tensile strength and tensile modulus, lightweight, high chemical resistance, good abrasion resistance, thermal insulation, high-temperature stability with relatively high stiffness, low thermal expansion, and electrical conductivity. However, carbon fiber is more expensive than the other types of fiber like basalt or glass fibers [178].

2.4.1.4 Glass Fiber

Glass fiber is a high-strength fiber (typical filament diameter of 3–20 μm) that is usually used to strengthen polymer composites, due to considerable characteristics such as great tensile strength and tensile module, non-corrosive, non-conductive, and non-magnetic properties [18,122]. The elastic modulus of glass fiber is approximately one-third of steel fiber. But, glass fiber elongates much more than carbon fiber before failure. Sensitivity to alkaline is the main drawback of glass fibers [178]. Sometimes silane-coupling agent is used to coat the surface of the glass fiber, to strengthen the interface between epoxy resin matrix and glass fiber [85,179].

2.4.1.5 Polyamide (Nylon) Fiber

Polyamide (nylon) synthetic fiber is an economically attractive material with generally outstanding properties. Because of the close chemical proximity to the resin of PC, nylon fiber can also be effective. It improves the impact resistance of polymer composites due to stretching and pulling out of the nylon fibers. However, only a small influence on tensile or bending strength was reported [122,136,180]. Nylon fiber also improves compressive strength and strain at yield point [181].

2.4.1.6 Polypropylene Fiber

Polypropylene fiber, a thermoplastic polymer fiber, due to some advantages including hydrophobic surface, good water, and alkali resistance, is produced in a variety of shapes and structures [178,182].

2.4.1.7 Coconut Fiber

Natural coconut fiber (coir) is produced from the fibrous husk (mesocarp) of the coconut from the coconut palm. The filament diameter of coconut fiber varies from 5 to 450 μm. It has a high lignin content and thus a low cellulose content, making fibers stiffer and tougher, resilient, strong and highly durable, high air porosity (voids appear from the dried out sieve cells), thermal hinder, biodegradable, and renewable source [72,183].

2.4.1.8 Sugar Cane Bagasse Fiber

Purifying process of sucrose from the sugar cane stalk remains a large amount of bagasse, containing both crushed rind and pith fibers (lignocellulosic fibers). After sifting and boiling in an alkaline solution, lignin removes from the waste bagasse and production process, followed by washing and drying. Incorporating bagasse fiber for reinforcement achieved lightweight and renewability composite, compared to pure synthetic materials [72].

2.4.1.9 Banana Fiber

It is a natural cellulosic fiber with quite remarkable mechanical properties produced from banana plants [72]. In addition to the fibers mentioned above, some other locally available natural fibers like bamboo and Tonkin cane, hemp rope, Chinese silver grass [184], singkut leaf fiber [93], and sisal fiber [185] have been implemented in PC reinforcement. Some of the physical, mechanical, and thermal properties of commonly used fibers in PC reinforcement are reviewed in Table 2.5.

2.4.2 Fiber Length

Fiber length is considered an important variable in PC performance. Although it may be possible to use unidirectional fibers to reinforce PC in specific applications, most of the chopped fibers are of interest to researchers. Chopped strand fibers are applied in PC reinforcement to randomly homogeneous dispersion of fibers in unhardened PC mixture by usual mixing procedures (according to ACI 544.1R-96) [167]. It can be seen from previous works that various fiber lengths from 0.1 [72] to 50 mm [186] were used in PC reinforcement.

2.4.3 Fiber Weight Content

Fiber amount is another important variable that affects reinforced PC properties. It is often expressed in weight content. However, some authors added fibers by volume [89]. Implementing fiber lower than the critical value causes a low contribution of fiber

TABLE 2.5
Physical, mechanical, and thermal characteristics of some synthesis fibers

Properties	Basalt fiber	Glass fiber	Carbon fiber	Steel fiber	Nylon fiber	Polypropylene
Density (kg/m^3)	2,700–6,800	2,500–2,700	1,750–2,270	7,800	1,500–5,000	910
Tensile strength (MPa)	3,480–4,840	2,000–3,500	3,500–5,000	345–1,200	900–965	550–700
Tensile modulus (GPa)	89–110	70–86	230–300	200	5–5.1	3.5–6.8
Poisson ratio	0.2	0.22	0.1	0.3	0.28	0.2
Elongation at break (%)	3.15	2.5–4	1.3–2	0.5–3.5	23–43	15–30
Linear CTE (10^{-6}/°K)	5.5–8	5–6	−0.1 to −1.3	11	7	11–13
Thermal conductivity (W/mK)	0.031–0.038	0.034–0.04	21–180	50.2	0.171–0.25	0.11–0.22
Application temperature (°C)	−260 to 982	−60 to 650	−50 to 700	—	−40 to 140	−160 to 130
Melting temperature (°C)	1,450	1,064–1,500	3,600–3,700	1,370–1,500	210–265	165

to the modulus and strain of PC. On the other hand, the large incorporation amount of fiber leads to weak resin-fiber interfacial bonding because of low enough resin to contain fibers [171] (see Figure 2.7). From the previous works, the fiber amount was investigated as an important variable and differed from 0.25 up to 6 wt%.

2.4.4 Effects of Fiber on Polymer Concrete Properties

As reported in the literature, fiber reinforcement can improve PC performance according to the utilized fiber type, length, and amount [72]. Interfacial bonding between polymer matrix and fibers due to reinforcement improves force transferring from the matrix to the fiber and gradually supports the entire load. Thus, due to stress transfer, the tensile strength of the reinforced PC increases. It is reported that fiber-reinforced PC had a more stable crack pattern than the unreinforced one [136].

Furthermore, fiber reinforcement increases PC toughness due to fiber bridging and fiber debonding toughening mechanisms. In fiber bridging toughening, some fibers debond but do not break under loading. Thus, fibers carry out stress, decrease the stress at the crack tip and prevent crack propagation.

The force transmitted to the matrix is transmitted by shear force to the fiber, under loading. The shear force eventually becomes so large that it breaks the fiber-matrix bonding. By increasing the applied force, cylindrical cracks in the contact surface

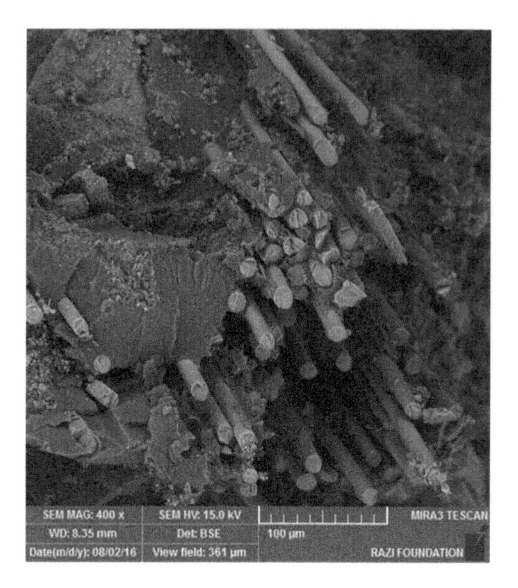

FIGURE 2.7 SEM micrograph from fracture surface of basalt fiber-reinforced PC [174]: the low amount of resin leads to weak matrix-fiber bonding

that originate from the matrix crack extend over the fiber. So fiber debonding occurs. Due to the variable strength of some fibers along their lengths, the fibers can be broken at some distance from the crack plane of the matrix, where the stress is very high. After failure, the composite usually shows the cracks with the fibers removed, called fiber pull-out. Breaking the fibers from the matrix leads to energy absorption and energy loss. After the separation and breakage of the first fiber bond, the portion of the energy decreases. Given that the failure energy has been reduced due to exposure to the previous fiber, it will no longer be able to separate the fiber and matrix, completely.

As a result, incorporating the fibers in the PC improves the toughness due to the fiber debonding mechanism. Also, the fibers act as fillers and fill the space and the bond between the cavities (see Figure 2.8). Therefore, introducing fiber to the PC can increase the mechanical properties [174].

Although the implementation of fiber improves the overall performance of PC, studies have shown that they may improve some properties and also reduce some others. As an example, glass fiber has a slight effect on the pre-cracking manner but improves the post-cracking response, which improves toughness and ductility, as well as tensile, flexural, and impact strengths [122]. The inclusion of chopped fibers, however, reduces the workability and applicability of PC [187], which can be attributed to the decreased flowability and difficulty in providing adequate finishing.

The effect of the fiber on the strength (compression and bending) of the PC is different from the elastic modulus. Carbon and glass fibers do not enhance the elastic modulus of compression [188]. Table 2.6 reviews the effects of fiber variables on PC properties.

FIGURE 2.8 SEM image of basalt fiber-reinforced PC: the fibers filled the voids [174]

2.5 NANOFILLER

Nanoparticles can improve the properties of PC just like polymer composite. Nanoparticles act as nanofiller in PC provide extensive contacts area and preventing the origination of a considerable number of subcritical microcavities and microcracks. The inclusion of nanofillers in PC can generate chemical interactions and new chemical bonds, which can improve mechanical properties.

2.5.1 Nanomaterial Types

Although the use of nanoparticles in PC will improve some of the properties, they should be considered in their economic aspects. The use of nanoparticles in concrete, for use with high amounts of concrete, is not economical. Therefore, cheaper nanoparticles should be used as much as possible. Different nanoparticles like nanoalumina, iron oxide, nanosilica, nanoclay, and carbon nanotube have been implemented to PCs, which will be reviewed in the following.

2.5.1.1 Nanoclay

Nanoclay or natural montmorillonite (MMT) is the most applied nanofiller in PC research until now. It is reported that different types of nanoclay were incorporated into PCs, such as natural montmorillonite, silane-modified Na^+-MMT, montmorillonite modified with methyl, tallow, bis-2-hydroxyethyl, and montmorillonite modified with dimethyl, dehydrogenated tallow 2-ethyl-hexyl [83].

It is proved that implementing nanoclay as a nanofiller in polymer nanocomposites improves some properties because of the high surface-to-volume ratio and extensive contact area [198,199]. Nanoclay provides good adhesion bonding with the polymer matrix and also prohibits portion movements of the polymer chains, and enhances the cross-linked density [83]. The incorporation of nanoclay on PC does not improve

TABLE 2.6
Effects of fiber variables on PC properties

References	Resin	Aggregate	Micro filler	Fiber and its variables	Studied properties	Brief findings
[174]	Epoxy	Siliceous sand, crushed basalt	Fly ash	Basalt fiber: 0–3.5 wt%, 6 mm length, 13 μm diameter	Compressive, Flexural, split tensile and impact strengths, residual strength in high temperatures	1. Inclusion 2 wt% fiber enhanced the compressive, flexural, and splitting tensile strengths up to 10%, 4.8%, and 35%, respectively 2. Implementing 3 wt% fiber increased the impact strength up to 4.15 times 3. Residual strengths of fiber-reinforced PC had less decrement than plain PC. Basalt fiber improved the thermal resistance
[75]	Epoxy, polyester	Ottawa sand, blasting sand		Glass fibers: 0–4 wt%, 13 mm length	Compressive, flexural, and split tensile strengths	1. Incorporating glass fibers increased the flexural and compressive strengths, toughness, and stiffness of PCs 2. Treatment of glass fibers with a silane-coupling agent enhanced both strength and stiffness
[168]	Epoxy	Sand		Steel fiber: 0–3.5 wt%, 15 mm length, 0.24 mm diameter	Flexural strength, creep	The inclusion of 3.5% fiber increased the flexural strength by 40%
[179]	Epoxy	Siliceous sand		Carbon fiber: 2 wt%, 6 mm length; Glass fiber: 1 wt%, 6 mm length	Compressive strength	1. Carbon and glass fibers improved compressive strength by 16% and 8.7%, respectively 2. Both carbon and glass fibers reinforcements do not enhance the compressive elastic modulus of PC 3. The values of Poisson's ratio didn't show a clear behavior 4. Failure behavior of the polymer matrix was altered to ductile failure

[189]	Polyester	Blasting sand and crushed sand		Glass fibers: 0–6 wt%	Mechanical and fracture properties	1. Implementing glass fiber improved the strength and toughness 2. Silane treatment of aggregate and fibers improved the flexural strength
[18]	Polyester	Blasting sand		Glass fibers: 0–6 wt%, 13 mm length	Compressive strength	1. Addition of 4 wt% fiber increased compressive strength up to 33% 2. Failure strain and toughness increased with the incorporation of fiber
[167]	Polyester	Gravel, sand		Steel fibers: 0–2 wt%, 30 mm length, 0.5 mm diameter	Compressive strength	1. Incorporating more than 1.3 wt% steel fibers increased the compressive strength from 80 to 100 MPa 2. Steel fibers increased the ductility of the PC which leads to a superior post-peak manner
[186]	Epoxy	Aggregate	Fly ash	Glass fibers: 0.5 wt%, 10–50 mm long; Polyester fiber: 0.5 wt%, 10–50 mm length; Metallic fiber: 2 and 7 wt%, 25 and 50 mm length; Cellulose fiber: 0.25 wt%	Compressive, flexural, and split tensile strengths	1. All types of fibers, except cellulose, improved the compressive strength 2. All types of fibers improved flexural strength. The maximum value was achieved for metallic fiber 3. Polyester, glass, and cellulose fibers increased the splitting tensile strength and the highest value was obtained by polyester fiber

(Continued)

References	Resin	Aggregate	Micro filler	Fiber and its variables	Studied properties	Brief findings
[190]	Polyester	Granite	Mineral filler	Copper coated stainless steel fibers: binder/ aggregate/filler = 1:8:1 (in weight), *L/d* ratio of 70	Static, dynamic, and thermal properties	1. Addition of steel fibers improved the properties of PC 2. Steel fiber enhanced the compressive strength of PC
[191]	Polyester	Blasting sand	—	Glass fiber: 0–6 wt%; carbon fiber: 0–6 wt%	Compressive and tensile strengths, damping ratio	1. Inclusion 6 wt% glass fibers increased compressive strength and the failure strain by 40% 2. Carbon fibers did not have any considerable influence on the compressive strength 3. Damping ratio enhanced with the introduction of glass and carbon fibers
[85]	Epoxy	Granite	—	Glass fiber: 1–5 wt%, 5–25 mm length	Vibration alleviating, damping ratio	1. Increasing fiber length first increased the damping ratio and later decreased that. The maximum damping value was achieved by 20 mm glass fibers 2. The least influential factor on damping ratio is fiber content. Increasing glass fiber amount enhanced the damping ratio 3. Flexibilizer dosage, length of the fiber, and epoxy amount have comparatively fewer affections on damping ratio 4. Granite mix is the critical variable for damping control, and the maximum damping is obtained by 16% resin, 5% glass fiber, and granite with ahigh ratio of fine aggregate

[188]	Epoxy	Siliceous sand	—	Carbon fiber: 2 wt%, 6 mm length; Glass fiber: 1 wt%, 6 mm length	Fracture properties	1. Carbon fiber and glass fiber improved fracture toughness by 29% and 13%, respectively 2. Implementing the fibers improved the TPM parameters and K_{Ic} by three times compared to OPC concrete
[72]	Epoxy	Siliceous sand	—	Coconut fiber: 2 wt%, 0.1–0.4 mm diameter; Sugar cane bagasse fiber: 2 wt%, 0.2–0.4 mm diameter; Banana fiber: 2 wt%, 0.154 mm diameter	Fracture mechanic and flexural strength	1. Both coconut and sugar cane bagasse fiber increased fracture toughness and fracture energy 2. Banana fiber did not improve fracture toughness. Only the fracture energy had an increment 3. Coconut fiber slightly increased the flexural properties of epoxy PC 4. Sugar cane bagasse and banana fiber reduced the flexural strength
[192]	Epoxy	River gravel	Fly ash	Cellulose fiber: 3 wt%, 10 μm length	Compressive, flexural, and split tensile strengths	1. Incorporating cellulose fibers decreased the mechanical properties of PC 2. Cellulose fiber reinforcement is not a suitable option for PC

(Continued)

References	Resin	Aggregate	Micro filler	Fiber and its variables	Studied properties	Brief findings
[193]	Epoxy	Siliceous sand	—	Carbon fiber: 2 wt%, 6 mm length; Glass fiber: 1 wt%, 6 mm length	Fracture energy	1. Due to its better mechanical properties, carbon fiber provided superior fracture properties than glass fiber 2. The evaluated fracture energy is lower than OPC concrete
[194]	Epoxy	Foundry sand		Recycled textile chopped fiber: 1 and 2 wt%, between 2 and 6 cm length	Compressive strength and flexural strength	1. The reinforced PC exhibited lower flexural and compressive properties 2. Textile fibers eliminated the signs of brittleness behavior of unreinforced PC
[136]	Polyester	Pea gravel, siliceous river sand	$CaCO_3$	Carbon fiber: 1 and 1.5 wt%, 6 mm length, 7 μm diameter; Nylon fiber: 1 and 1.5 wt%, 6 mm length, 23 μm diameter	Fracture properties, compressive strength, and flexural strength	1. Nylon and carbon fibers improved the compressive strength of PC up to 15.6% and 4.0%, respectively 2. Carbon fiber provided higher flexural strength than nylon fiber, while carbon fiber provided higher tensile strength 3. Carbon fiber provided better mechanical properties and fracture toughness than nylon fiber
[71]	Polyester and epoxy	Siliceous sand	—	Glass fiber: 0–4 wt%, 6 mm length	Compressive strength, flexural strength, and Interfacial shear	1. Inclusion of the m fibers up to 2 wt% increased the flexural strength 2. Decreasing fiber up to 0 wt% increased the compressive strength 3. Increasing the fiber content decreased the interfacial shear strength

[11]	Polyester	Siliceous sand	—	Glass fiber: 1.5 wt%, 12.5–18 mm length	Modulus of rupture, fracture toughness	1. The rupture modulus of PC, including 20% polyester and about 79% sand, was about 20 MPa, and the fracture toughness was obtained about 1.2 $MNm^{-3/2}$ 2. Incorporation of 1.5 wt% glass fibers enhanced the modulus of rupture and fracture toughness by about 20% and 55%, respectively 3. Glass fiber improved the strength by enhancing the required force for deformation and enhanced the toughness by enhancing the required energyfor crack propagation
[89]	Polyester	Marble	—	Polypropylene fiber: 0.1–0.3% by volume and radiation doses	Compressive properties	1. Incorporating PP fibers, increased compressive strength up to 24% 2. Enhancing the fiber concentration increased the compressive strain 3. By enhancing the PP fiber concentration, the compressive modulus decreased
[195]	Epoxy	Ordinary sand	P.O 42.5 cement	Steel fiber: 0–6 wt%	Flexural and compressive strengths	1. Small amount of steel fiber did not improve the PC strength 2. 3 wt% steel fiber increased the bending strength by 26% and the compressive strength by 6%
[184]	Polyester	Desert sand	—	Bamboo cane: 10 and 35 mm diameter: Tonkin cane: 10 and 15–20 mm diameter; Chinese silver grass: 5–15 mm diameter	Fracture and load-bearing	1. The residual load-bearing capacity of PC reinforced by bamboo and Tonkin canes was equal to or higher than unreinforced PC 2. Chinese silver grass was not appropriate due to weak bonding because of the alternating leaves on the stems 3. Tonkin canes increased the load-bearing potential of PC at elevated temperatures

(Continued)

References	Resin	Aggregate	Micro filler	Fiber and its variables	Studied properties	Brief findings
[196]	Polyester	Silica sand	—	Recycled tire fiber: 0.3%, 0.6%, 0.9%, and 1.2% by volume; gamma irradiation (50 and 100 kGy) modification	Compressive and flexural strengths	1. Non-irradiated fibers improved compressive and flexural strengths by 41% and 37%, respectively, and also improved compressive and flexural deformations by 25% and 20%, respectively 2. Adding irradiated fibers improved compressive and flexural strengths by 26% and 48%, respectively at 100 kGy, and also improved compressive and flexural deformations by 52% and 29%, respectively, at 50 kGy 3. Ionizing radiation caused cross-linking and scission of polymer chains, and the morphological conflict changes as well as for the degree of crystallinity
[197]	Epoxy	Andesite gravel, silica sand	—	Carbon fiber: up to 15 wt%, 3 mm length,	Flexural Strength, vibration damping	1. Addition of 12% fibers increased the flexural strength up to 17.9 MPa 2. PC with 3% fiber decreased vibration acceleration deviation up to 93.5% in 0.005 seconds

all properties. Nanoclay improved the compressive, flexural, and impact strengths as well as the thermal resistance of the PC but reduced the tensile strength [174].

2.5.1.2 Nanoalumina (Al_2O_3)

Alumina nanoparticles (ANPs) are simply nanoscale aluminum oxide (Al_2O_3) particles. The incorporating ANPs improved the tensile strength, modulus of elasticity, fracture toughness, fracture energy, ductility, bond strength, fire reaction properties, resistance to fatigue crack propagation, and durability of PC. However, a slight decrease in compressive and flexural strengths was reported [200–202].

2.5.1.3 Iron Oxide (Fe_2O_3)

It is reported that the introduction of iron oxide nanofiller to the PC increased the modulus of elasticity, fracture toughness, fracture energy, and fire reaction properties. However, no improvement in flexural and compressive strengths was observed [201,203].

2.5.1.4 Nanosilica (SiO_2)

Silica nanoparticles (SNPs) are non-toxic, non-reactant, non-polluting with flocculent and mesh microstructure. They can improve the basic features of the polymer composite, such as strength and elastic properties, due to the formation of three-dimensional network structures with polymer chains [195]. Also, nanosilica increase the compressive strength, impact strength, and Poisson's ratio as well as stiffness of PC. However, nanosilica decreased the elastic modulus, tensile, and bond strength [200,202].

2.5.1.5 Carbon Nanotube

Carbon nanotubes (CNTs) have remarkable tensile strength, elastic modulus, toughness, and failure strain, as well as a significant aspect ratio. Uniformly dispersed CNTs improve the flexural and tensile strengths, flexural and shear failure strains, flexural and shear toughness, bond strength, ductility, overlays cohesion strength, fatigue resistance, fracture toughness, thermal stability, and electrical conductivity of PC [200,202,204–206].

2.5.2 Nanomaterial Weight Content

Researchers introduced various nanoparticles in PC in the range of 0.5% up to 10% of resin weight content. It was found that the best results were obtained by adding 2–3 wt% nanoparticles in resin in the case of PCs. The inclusion of a high dosage of nanoparticles in PC composites has an adverse effect on properties. It's due to the considerable increase in viscosity and decreased PC flowability at a high dosage of nanoparticles. Entrapping air due to high viscosity decrease the mechanical properties of PC. Also, the large incorporation amount of nanofiller results in inhomogeneous dispersion and agglomeration of nanoparticles in polymer binder and reduces properties. From the literature, sometimes a little weight content of nanoparticles as low as 0.5 wt% increases the properties more than 100% [205]. Choosing a suitable range of nanoparticles is contingent on the expected properties of the nanocomposite and should be selected accordingly [207].

2.5.3 Nanocomposites Preparation Methods

Nanoparticles, due to their great contact area, generate a high level of shear friction, which makes it challenging to obtain a homogeneous dispersion. The desirable effect of nanoparticles is influenced by the level of dispersion [208–210]. An appropriate dispersion of nanomaterials can be defined as separating those particles into individual particles within the polymer binder. As reported, entangled, agglomerated, and concentrated arrangement of nanofillers in polymer binder resulted in a dramatic reduction of properties, as well as a significant increase in viscosity [211–214]. Nanoparticle dispersion requires providing sufficient energy to de-agglomerate the particles. Nanoparticles dispersion techniques in PC followed in literature are summarized by ultrasonication via bath or probe, magnetic stirring, shear mixing, or by combining several techniques.

Both probe sonication and shear mixing are efficient methods with high shear energy. However, they are often expensive and require careful application. Some authors implemented a probe sonication approach after shear mixing for achieving the homogenous dispersion of the nanoparticles in PC [174,207].

Stirring is a cheaper dispersion method but provides significantly less shear energy. Therefore, it is not considered an effective homogenization technique unless it couples with functionalization or sonication methods to maintain particle suspension [202].

Ultrasonication bath, a relatively less expensive dispersion technique, enhances dispersion quality by producing microbubbles within the resin that release energy and prohibit nanofiller aggregation, therefore, leading to a homogeneous distribution [200].

Because of having a high aspect ratio and quite formidable van der Waals forces, carbon nanotubes, and the other high aspect ratio nanomaterials are more difficult to uniformly distribute than the other nanoparticles [215]. MWCNTs could be surface functionalized with functional groups to generate repelling forces to facilitate distribution [206,210,215]. The dispersion process of multi-malled carbon nanotubes, aluminum nanoparticles, and silica nanoparticles is accomplished by magnetically stirring at about 800 rpm for two hours. Afterward sonication process is conducted at 60°C for about two hours. Degassing by the sonicator hot bath option was conducted to remove the air bubble within the resin [200].

For dispersion of nanoclay in epoxy PC, nanoparticles mixed for one hour at 2,000 rpm by implementing a two-propelled mechanical stirrer and then probe sonicator with different sonication output powers of 100, 200, and 400 W for 30 minutes and 100 W for 0.25, 0.5, 1, 2, and 4 hours were implemented to evaluate the appropriate sonication power and time for obtaining the highest d-spacing of nanolayers. As a result of XRD study, the maximum of d-spacing was almost gained after 15 minutes of sonication. Finally, optimum sonication was achieved at 100 W for 30 minutes, due to better degassing and also prohibiting overheating [207]. Sometimes an ice-bath cooling system is implemented for prohibiting the mixture during probe sonication from overheating [174,207].

2.5.4 Effects of Nanofiller on PC Properties

Nanomaterials provide better packing density as well as the ability to alter the cross-linking density and improve stress transfer similar to micro fillers effects [202]. The introduction of 1 wt% nanoclay improved the compressive strength of basalt fiber-reinforced PC by 7%. Maximum flexural strength was achieved by 2 wt% nanoclay and was 27%, higher than unreinforced PC. However, the splitting tensile strength decreased. Experimental investigation of the Izod impact strength exhibited that nanoclay enhanced the impact strength up to 2.6 times. Thus, clay nanoparticle has the maximum influence on the impact strength of PC [174]. Also, nanoclay increases the toughness of PC as well as the other polymer-based composites and generates rough fracture surfaces, as can be observed in the literature [174,207]. It was determined that the inclusion of nanoclay on PC enhances the compressive strength up to 15.2%, however, tensile and flexural strengths do not improve [207].

Clay nanoparticles also improve the thermal stability of PCs considerably. In MMT-reinforced unsaturated polyester PC, elevating the temperature from 25°C to 65°C declined the compressive strength, modulus of elasticity, splitting tensile strength, and flexural strength. But the decrease in properties was smaller than that of unreinforced PC [83]. Experimental investigation on effects of elevated temperatures of 25°C, 50°C, 100°C, 150°C, 200°C, and 250°C on the mechanical strength presented that the compressive and flexural strengths of PC, including 2 wt% nanoclay at elevated temperatures, are higher than PC without nanoparticles. However, splitting tensile strength of clay-reinforced PC has similar behavior to unreinforced one [174].

It is reported that a sudden drop in PC commonly takes place as the impact target breaks the specimen and starts to penetrate it. Adding MWCNTs [204] decreases sudden drop in impact resistance which reflects the ability of fiber-reinforced PC to increase the residual strength after the peak force, dissipate and absorb energy compared with unreinforced PC, as illustrated in Figure 2.9. The addition of 0.5%, 1.0%, and 1.5% MWCNTs decreased the creep compliance of PC at the failure by 68%, 7%, and 35%, respectively. Table 2.7 reviews and summarizes the effects of the addition of nanomaterials on PC performance.

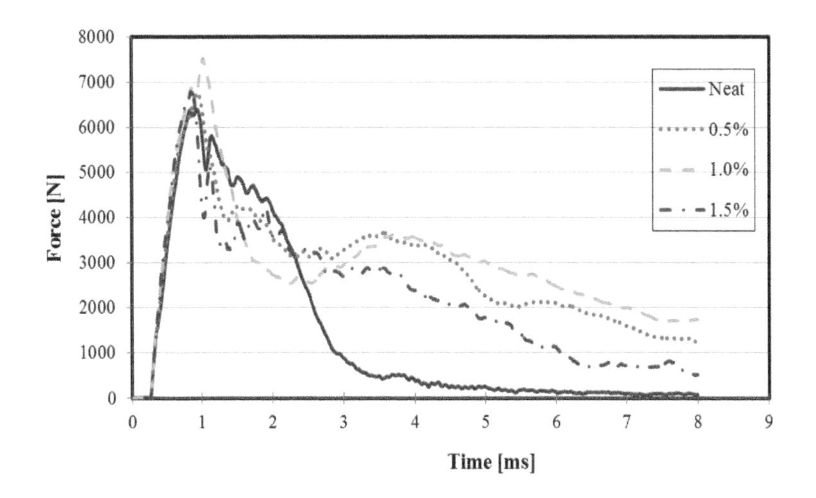

FIGURE 2.9 The effect of MWCNTs on residual strength after the peak force and energy absorption in low-velocity impact test [204]

TABLE 2.7
Effects of nanomaterial variables on PC properties

References	Resin	Aggregate/ micro filler	Fiber	Nanomaterial and its variables	Studied properties	Results
[83]	Unsaturated polyester	Pea gravel and siliceous river sand/$CaCO_3$	—	Three kinds of nano-montmorillonite: non-treated Na^+-MMT, montmorillonite modified with methyl, tallow, bis-2-hydroxyethyl, montmorillonite modified with dimethyl, dehydrogenated tallow 2-ethyl-hexyl, content: 2, 5, 8, and 10 wt%	Flexural and splitting tensile strengths	1. Incorporation of nanoparticles increased flexural and split tensile strengths 2. The strength and elastic modulus of the PC were increased by using exfoliated MMT-UP nanocomposite 3. The strength and elastic modulus were found to be positively correlated with the tensile strength and tensile modulus of the MMT-UP nanocomposite
[174]	Epoxy	Crushed basalt and silica sand/fly ash	Basalt fiber 2 wt%, 6 mm length, 13 μm diameter	Montmorillonite nanoclay, content: 1, 2, 3, and 4 wt%	Compressive flexural, splitting tensile, and impact strengths, thermal stability	1. Addition of nanoclay improved the compressive, flexural, and impact strengths. However, the splitting tensile strength decreased 2. Adding nanoclay led to the maximum increase on the impact strength and then on the flexural strength 3. Residual compressive and flexural strength at different temperatures for nanoclay reinforced PC is higher than PC without nanoclay. Therefore, the addition of nanoclay to BFRPC enhanced its thermal stability

[207]	Epoxy	Coarse aggregates, siliceous sand	E-glass fiber, 0.5 wt%, 6 mm length	Montmorillonite nanoclay modified with a methyl, tallow, bis-2-hydroxyethyl material, content: 1–5 wt%	Compressive, tensile, and flexural strengths, fracture toughness	1. Incorporation of nanoclay did not increase tensile and flexural strengths 2. Addition of nanoclay increased the compression strength and fracture toughness by 15.2% and 7.6%, respectively
[204]	Epoxy	Bauxite-based aggregate/ crystalline silica (quartz) and ceramic microspheres	—	MWCNTs, content: 0.5, 1.0, and 1.5 wt%	Compressive, flexural, shear, creep, and bond strengths (pull-off and slant shear tests), the fatigue strength of PC overlays under cyclic traffic loads, and dynamic response	1. The introduction of MWCNTs to the PC improved the flexural strength, flexural failure strain, shear failure strain, flexural and shear toughness, overlays cohesion strength, the adhesion strength between PC overlays, and steel, resistance to damage propagation under cyclic loading. However, it had no remarkable influence on shear strength and reduced the penetration velocity and flowability. Also, fast failure in the creep test was observed 2. From TGA and DSC analysis adding MWCNTs enhanced thermal stability and maximum decomposition temperature
[70]	Epoxy	Siliceous sand	—	Nano-Al_2O_3 and nano-Fe_2O_3 content: 3, 5, 7, and 10 wt%	Compressive strength and flexural strengths	1. Addition of nanoparticles did not improve flexural and compressive strength. Lower concentrations of nanoparticles caused a slight decrease in properties 2. Introduction of nanoparticles provided a remarkable increment on the flexural and compressive elastic modulus of the PC

(Continued)

References	Resin	Aggregate/ micro filler	Fiber	Nanomaterial and its variables	Studied properties	Results
[203]	Epoxy	Siliceous sand	—	Nano-Al_2O_3 and nano-Fe_2O_3 content: 3, 5, 7, and 10 wt%	Fracture properties	1. Addition of both Al_2O_3 and Fe_2O_3 increased the elastic modulus, fracture toughness, and fracture energy 2. 3.0 wt% of both nanoparticles provided the best results as a toughening mechanism 3. Increasing the amount of the nanoparticles reduced properties, but values are still greater than unreinforced PC
[201]	Epoxy	River sand, siliceous sand/spherical micro alumina particles	—	Nano-Al_2O_3 content: 5–10 wt%	Compressive strength, flexural strength, fire reaction properties	1. Inclusion of Al_2O_3, either in the dry state or into a dispersion median, enhanced the compressive and flexural strength. The incorporation of 3.9 wt% nano-Al_2O_3 powder caused a slight improvement in flexural strength. The exact nanoparticles content in a dispersion median led to a decrease in these properties 2. Addition of 3.9 wt% Al_2O_3 powder caused quite an improvement in fire reaction properties of PC, especially about SEA and HRR indexes. Size reduction of Al_2O_3 particles from micro to nano improved flame retardancy. Nano dispersion of Al_2O_3 did not improve the fire reaction of PC

[202]	Two types of epoxies	Crystalline silica (quartz) and ceramic powder	—	Alumina nanoparticles (ANPs), silica nanoparticles (SNPs), pristine and carboxyl multi-walled carbon nanotubes, content: 1–4 wt%	Compressive, flexural, fracture toughness, tension, flowability, slant shear, and fatigue strength alongside electrical conductivity monitoring	1. All applied nanomaterials improved Poisson's ratio and the bond strength to steel substrates. However, the elastic modulus decreased 2. Addition of 2 wt% P-MWCNTs increased fatigue life by 55% while 2 wt% ANPs decreased it by 50% 3. P-MWCNTs, COOH-MWCNTs, and ANPs improved the fracture toughness 4. 2 wt% P-MWCNTs improved the conductivity by three orders of magnitude 5. The best ductility and toughness were achieved for COOH-MWCNTs, then ANPs and P-MWCNTs
[200]	Epoxy	Crystalline silica (quartz) and ceramic powder	—	Multi-walled carbon nanotubes (MWCNTs), aluminum nanoparticles (ANPs), silica nanoparticles (SNPs), content: 0.5, 1, and 2 wt%	Bond strength to the steel substrate	1. MWCNTs at 0.5, 1.0, and 2.0 wt% increased bond strength by 7%, similarly 2. Incorporation of 0.5, 1.0 and 2.0 wt% ANPs increased bond strength by 20%, 23%, and 51%, respectively 3. Implementing the 0.5 and 1 wt% SNPs enhanced bond strength by 7 and 18%. However, a sharp drop of 41% at 2 wt% SNPs was observed
[195]	Epoxy	Ordinary sand/P.O 42.5 cement	PP fiber (added after nanofiller)	Silica Nanoparticles (SiO_2), content: 0, 2, 4, 6, and 8 wt%	Tensile strength, impact strength	1. SiO_2 has a toughening effect on PC 2. Addition of SiO_2 nanoparticles, increased impact, and tensile strengths of PC. The best tensile and impact strengths were achieved by 6 and 2 wt% SiO_2 3. Optimum properties were obtained by 2 wt% SiO_2 (1.24 J/m, 1.43MPa)

(Continued)

References	Resin	Aggregate/ micro filler	Fiber	Nanomaterial and its variables	Studied properties	Results
[216]	Epoxy	Crystalline silica (quartz) and ceramic powder	—	Alumina nanoparticles (ANPs), content 0.5, 1.0, 2.0, and 3.0 wt%	Tensile strength, tensile failure strain, and fracture toughness (K_{Ic}, GIC, and J_{Ic}), flowability	1. Inclusion of 3.0 wt% ANPs increased the tensile failure strain from 2.56% (for the neat PC) to 4.89% 2. ANPs significantly improved ductility by 60.6% and fracture toughness by 131.8%. It also improved the toughness, strain hardening, and fracture energy dissipation 3. 2–3 wt% ANPS decreases the flowability, but it is still relatively flowable concrete 4. No significant improvement of tensile strength was observed
[205]	Epoxy	Crystalline silica (quartz) and bauxite-based aggregate	—	Multi-walled carbon nano tubes (MWCNTs), aluminum nanoparticles (ANPs), content: 0.5, 1, and 1.5 wt%	Compressive, flexural, impact and tension strengths, fracture toughness	1. Incorporating MWCNTs and ANPs considerably improved the mechanical strength, ductility, and fracture toughness of PC 2. Incorporation of pristine MWCNTs considerably increased ductility and altered the electrical conductivity of PC, enabling a conductive material 3. Adding COOH- MWCNTs as low as 0.5 wt% doubled the fracture toughness of PC 4. Introducing ANPs improved ductility and fracture toughness of PC significantly

[206]	Epoxy	Ceramic microsphere powder and crystalline silica (quartz)	—	Pristine and carboxyl functionalized MWCNTs, content: fixed 2 wt% mixes of them	Tensile strength, cross-linking density	1. Addition of pristine and functionalized MWCNTs led to unprecedented ductility levels with a failure strain of 5.5% 2. Increasing the COOH functional group content increased polymer cross-linking 3. P-MWCNTs prevented the increase in cross-linking and thus improved ductility 4. PC with F-MWCNTs only reaches high tensile strength of 15.4 MPa
[217]	Epoxy	Ceramic microsphere powder and crystalline silica (quartz)	—	Pristine and carboxyl functionalized MWCNTs, content: fixed 2 wt% mixes of them	Tensile strength, ductility, flowability	1. COOH functionalization maximizes PC ductility up to 5.5% failure strains and improves toughness by 184%. An optimal COOH content for maximizing PC ductility is in the range of 0.001–0.018 wt% 2. MWCNTs reduced the flowability of PC up to 23% due to an increase of viscosity at the high content of MWCNTs 3. COOH amount is directly proportional to PC tensile strength. Tensile strength of the PC, including only COOH-MWCNTs, increase by 30%

REFERENCES

[1] R. Bedi, R. Chandra, Reviewing some properties of polymer concrete, *Indian Concr. J.* 88 (2014) 47–68.

[2] G. Odian, *Principles of Polymerization*, 4th ed., A John Wiley & Sons, INC, Staten Island, New York, 2004.

[3] M. Frigione, Concrete with polymers, in: *Eco-Efficient Concr.*, Woodhead Publishing Limited, 2013: pp. 386–436. https://doi.org/10.1533/9780857098993.3.386.

[4] K.B. Choi, N.J. Jin, Y.S. Lee, K.S. Yeon, Prediction of compressive strength of unsaturated polyester resin based polymer concrete using maturity method, *J. Korean Soc. Agric. Eng.* 59 (2017) 19–27.

[5] J.T. San-José, I.J. Vegas, Moisés Frías, Mechanical expectations of a high performance concrete based on a polymer binder and reinforced with non-metallic rebars, *Constr. Build. Mater.* 22 (2008) 2031–2041. https://doi.org/10.1016/j.conbuildmat.2007.08.001.

[6] A.S. Momtazi, R. Kohani Khoshkbijari, S. Sabagh Mogharab, Polymers in concrete: Applications and specifications, *Eur. Online J. Nat. Soc. Sci.* 4 (2015) 62–72.

[7] C. Kiruthika, S. Lavanya Prabha, M. Neelamegam, Different aspects of polyester polymer concrete for sustainable construction, *Mater. Today Proc.* 43 (2020) 1622–1625. https://doi.org/10.1016/j.matpr.2020.09.766.

[8] E. Hwang, J. Kim, J.H. Yeon, Characteristics of polyester polymer concrete using spherical aggregates from industrial by-products, *J. Appl. Polym. Sci.* 129 (2013) 2905–2912. https://doi.org/10.1002/app.39025.

[9] K.-S. Yeon, N.J. Jin, J.H. Yeon, Effect of methyl methacrylate monomer on properties of unsaturated polyester resin-based polymer concrete, in: M.M.R. Taha (Ed.), *Int. Congr. Polym. Concr. (ICPIC)*, Springer, Cham, Washington, DC, 2018: pp. 165–171. https://doi.org/https://doi.org/10.1007/978-3-319-78175-4_19.

[10] K. Ostad-ali-askari, V.P. Singh, Polymer concrete, *Int. J. Hydrol.* 2 (2018) 630–635. https://doi.org/10.15406/ijh.2018.02.00135.

[11] R. Gri, A. Ball, An assessment of the properties and degradation behaviour of glass-fibre-reinforced polyester polymer concrete, *Compos. Sci. Technol.* 60 (2000) 2747–2753.

[12] M.C.S. Ribeiro, C.M.L. Tavares, A.J.M. Ferreira, Chemical resistance of epoxy and polyester polymer concrete to acids and salts, *J. Polym. Eng.* 22 (2011) 27–44. https://doi.org/https://doi.org/10.1515/POLYENG.2002.22.1.27.

[13] W. Lokuge, T. Aravinthan, Effect of fly ash on the behaviour of polymer concrete with different types of resin, *Mater. Des.* 51 (2013) 175–181. https://doi.org/10.1016/j.matdes.2013.03.078.

[14] N. Ahn, Moisture sensitivity of polyester and acrylic polymer concretes with metallic monomer powders, *J. Appl. Polym. Sci.* 107 (2008) 319–323. https://doi.org/10.1002/app.

[15] F. Cakir, Effect of catalysts amount on mechanical properties of polymer concrete, *Chall. J. Concr. Res. Lett.* 11 (2020) 46–52. https://doi.org/10.20528/cjcrl.2020.02.001.

[16] J.M. Miranda Vidales, L. Narváez Hernández, J.I. Tapia López, E.E. Martínez Flores, L.S. Hernández, Polymer mortars prepared using a polymeric resin and particles obtained from waste pet bottle, *Constr. Build. Mater.* 65 (2014) 376–383. https://doi.org/10.1016/j.conbuildmat.2014.04.114.

[17] A. Sharma, M.I. Sayyed, O. Agar, M.R. Kaçal, H. Polat, F. Akman, Photon-shielding performance of bismuth oxychloride-filled polyester concretes, *Mater. Chem. Phys.* 241 (2020) 122330. https://doi.org/10.1016/j.matchemphys.2019.122330.

[18] B.S. Mebarkia, A. Member, Compressive behavior of glass-fiber-reinforced polymer concrete, *J. Mater. Civ. Eng.* 4 (1992) 91–105.

[19] M.E. Tawfik, S.B. Eskander, Polymer concrete from marble wastes and recycled poly(ethylene terephthalate), *J. Elastomers Plast.* 38 (2006) 65–79. https://doi.org/10.1177/0095244306055569.

[20] S. Mebarkia, C. Vipulanandan, Mechanical properties and water diffusion in polyester polymer concrete, *J. Eng. Mech.* 121 (1995) 1359–1365. https://doi.org/https://doi.org/10.1061/(ASCE)0733-9399(1995)121:12(1359).
[21] P. Amuthakkannan, V. Manikandan, J.T.W. Jappes, M. Uthayakumar, Effect of fiber length and fiber content on mechanical properties of short basalt fiber, *Mater. Phys. Mech.* 16 (2013) 107–117.
[22] K.S. Rebeiz, D.W. Fowler, D.R. Paul, Polymer concrete and polymer mortar using resins based on recycled poly(ethylene terephthalate), *J. Appl. Polym. Sci.* 44 (1992) 1649–1655. https://doi.org/10.1002/app.1992.070440919.
[23] X. Xie, X. Zhang, Y. Jin, W. Tian, Research progress of epoxy resin concrete, in: *IOP Conf. Ser. Earth Environ. Sci.* 186, 2018. https://doi.org/10.1088/1755-1315/186/2/012038.
[24] M. Nodehi, Epoxy, polyester and vinyl ester based polymer concrete: A review, *Innov . Infrastruct. Solut.* 7 (2022) 1–24. https://doi.org/10.1007/s41062-021-00661-3.
[25] W. Ferdous, A. Manalo, T. Aravinthan, G. Van Erp, Properties of epoxy polymer concrete matrix: Effect of resin-to-filler ratio and determination of optimal mix for composite railway sleepers, *Constr. Build. Mater.* 124 (2016) 287–300. https://doi.org/https://doi.org/10.1016/j.conbuildmat.2016.07.111.
[26] P. Mani, A.K. Gupta, S. Krishnamoorthy, Comparative study of epoxy and polyester resin-based polymer concretes, *Int. J. Adhes. Adhes.* 7 (1987) 157–163. https://doi.org/10.1016/0143-7496(87)90071-6.
[27] M. Muthukumar, D. Mohan, Studies on furan polymer concrete, *J. Polym. Res.* 12 (2005) 231–241. https://doi.org/10.1007/s10965-004-3206-7.
[28] R. Kumar, A review on epoxy and polyester based polymer concrete and exploration of polyfurfuryl alcohol as polymer concrete, 2016 (2016) 7249743.
[29] D.A. Hausmann, A. Lander, Products made with furan resin, US3301278A, 1967.
[30] R. Katiyar, S. Shukla, Studies on furan polymer concrete, *Int. Res. J. Eng. Technol.* 4 (2017) 721–727.
[31] M. Muthukumar, D. Mohan, Studies on polymer concretes based on optimized aggregate mix proportion, *Eur. Polym. J.* 40 (2004) 2167–2177. https://doi.org/10.1016/j.eurpolymj.2004.05.004.
[32] S.-W. Son, J.H. Yeon, Mechanical properties of acrylic polymer concrete containing methacrylic acid as an additive, *Constr. Build. Mater.* 37 (2012) 669–679. https://doi.org/https://doi.org/10.1016/j.conbuildmat.2012.07.093.
[33] Y. Ohama, Recent progress in concrete-polymer composites, *Adv. Cem. Based Mater.* 5 (1997) 31–40. https://doi.org/10.1016/S1065-7355(96)00005-3.
[34] K. Yeon, J. Cha, J.H. Yeon, I.B. Seung, Effects of TMPTMA and silane on the compressive strength of low-temperature cured acrylic polymer concrete, *J. Appl. Polym. Sci.* 131 (2014) 1–8. https://doi.org/10.1002/app.40939.
[35] K.-S. Yeon, J.H. Yeon, Y.-S. Choi, S.-H. Min, Deformation behavior of acrylic polymer concrete: Effects of methacrylic acid and curing temperature, *Constr. Build. Mater.* 63 (2014) 125–131. https://doi.org/https://doi.org/10.1016/j.conbuildmat.2014.04.051.
[36] M. Miller, *Polymers in Cementitious Materials*, Smithers Rapra Publishing, UK, 2005.
[37] V.K. Raina, *Concrete Bridges: Inspection, Repair, Strengthening, Testing and Load Capacity Evaluation*, McGraw-Hill, New York, 1996.
[38] N.J. Jin, K.S. Yeon, S.H. Min, J. Yeon, Using the maturity method in predicting the compressive strength of vinyl ester polymer concrete at an early age, *Adv. Mater. Sci. Eng.* 2017 (2017) 1–12. https://doi.org/10.1155/2017/4546732.
[39] H. Huang, H. Pang, J. Huang, P. Yu, J. Li, M. Lu, B. Liao, Influence of hard segment content and soft segment length on the microphase structure and mechanical performance of polyurethane- based polymer concrete, *Constr. Build. Mater.* 284 (2021) 122388. https://doi.org/10.1016/j.conbuildmat.2021.122388.

[40] M.H. Niaki, M.G. Ahangari, A. Fereidoon, Mechanical properties of reinforced polymer concrete with three types of resin systems, *Proc. Inst. Civ. Eng. Constr. Mater.* 0 (2022) 1–24. https://doi.org/10.1680/jcoma.21.00060.

[41] S. Ibrahim Haruna, H. Zhu, W. Jiang, J. Shao, Evaluation of impact resistance properties of polyurethane-based polymer concrete for the repair of runway subjected to repeated drop-weight impact test, *Constr. Build. Mater.* 309 (2021) 125152. https://doi.org/https://doi.org/10.1016/j.conbuildmat.2021.125152.

[42] H. Tokushige, M. Kawakami, Y. Kurimoto, H. Yamauchi, T. Sasaki, Porous polymer concrete using polyurethane resin and chipped aggregates made of wood wastes, in: *PRO 41 Int. RILEM Symp. Environ. Mater. Syst. Sustain. Dev.*, RILEM Publications, 2005: p. 321.

[43] M.C. Bignozzi, F. Sandrolini, E. Franzoni, New polymer mortars based on unsaturated polyester-polyurethane interpenetrating polymer networks, in: *Tenth Int. Congr. Polym. Concr.*, The University of Texas at Austin, Austin, TX, 2001: p. 12.

[44] R. Drochytka, J. Hodul, Experimental verification of use of secondary raw materials as fillers in epoxy polymer concrete, in: M.M.R. Taha (Ed.), *Int. Congr. Polym. Concr. (ICPIC)*, Springer, Cham, Washington, DC, 2018: pp. 135–141. https://doi.org/https://doi.org/10.1007/978-3-319-78175-4_15.

[45] A. Hugo, L. Scelsi, A. Hodzic, F.R. Jones, Development of recycled polymer composites for structural applications, *Plast. Rubber Compos.* 40 (2011) 317–323. https://doi.org/10.1179/1743289810Y.0000000008.

[46] K.S. Rebeiz, Time-temperature properties of polymer concrete using recycled PET, *Cem. Concr. Compos.* 17 (1995) 119–124. https://doi.org/10.1016/0958-9465(94)00004-I.

[47] M. Antonio, G. Jurumenha, J. Marciano, Fracture mechanics of polymer mortar made with recycled raw materials, *Mater. Res.* 13 (2010) 475–478.

[48] F. Mahdi, A.A. Khan, H. Abbas, Physiochemical properties of polymer mortar composites using resins derived from post-consumer PET bottles, *Cem. Concr. Compos.* 29 (2007) 241–248. https://doi.org/https://doi.org/10.1016/j.cemconcomp.2006.11.009.

[49] F. Mahdi, H. Abbas, A.A. Khan, Strength characteristics of polymer mortar and concrete using different compositions of resins derived from post-consumer PET bottles, *Constr. Build. Mater.* 24 (2010) 25–36. https://doi.org/http://dx.doi.org/10.1016/j.conbuildmat.2009.08.006.

[50] A.S. Benosman, M. Mouli, H. Taibi, M. Belbachir, Mineralogical study of polymer-mortar composites with PET polymer by means of spectroscopic analyses, *Mater. Sci. Appl.* 3 (2012) 139–150. https://doi.org/10.4236/msa.2012.33022.

[51] K.P. Singh, R. Bedi, A. Bakshi, Effect of silane treatment on the mechanical properties of polymer concrete made with recycled pet resin and recycled aggregates, in: *UKIERI Concr. Congr.*, Jalandhar, 2019: pp. 1–14.

[52] K.S. Rebeiz, D.W. Fowler, D.R. Paul, Recycling plastics in polymer concrete systems for engineering applications, *Polym. Plast. Technol. Eng.* 30 (1991) 809–825. https://doi.org/10.1080/03602559108021008.

[53] C. Meran, O. Ozturk, M. Yuksel, Examination of the possibility of recycling and utilizing recycled polyethylene and polypropylene, *Mater. Des.* 29 (2008) 701–705. https://doi.org/10.1016/j.matdes.2007.02.007.

[54] N. Choi, Y. Ohama, *Basic Properties of New Polymer Mortars Using Waste Expanded Polystyrene Solution-Based Binders*, 2003. https://doi.org/10.1515/POLYENG.2003.23.5.369.

[55] K. Chung, Y. Hong, Weathering properties of elastic rubber concrete comprising waste tire solution, *Polym. Eng. Sci.* 49 (2009) 794–798. https://doi.org/10.1002/pen.

[56] I. Roh, K. Jung, S. Chang, Y. Cho, Characterization of compliant polymer concretes for rapid repair of runways, *Constr. Build. Mater.* 78 (2015) 77–84. https://doi.org/10.1016/j.conbuildmat.2014.12.121.

[57] R.C. Prusinski, J.R. Prusinski, W.S. Wahby, Thermoplastic polymer concrete, in: J.B. de Aguiar, S. Jalali, A. Camoes, R.M. Ferreira (Eds.), *Int. Congr. Polym. Concr. (ICPIC)*, Funchal-Maderia, Portugal, 2010.
[58] W. Ferdous, A. Manalo, H.S. Wong, R. Abousnina, O.S. Alajarmeh, Y. Zhuge, P. Schubel, Optimal design for epoxy polymer concrete based on mechanical properties and durability aspects, *Constr. Build. Mater.* 232 (2020) 117229. https://doi.org/10.1016/j.conbuildmat.2019.117229.
[59] M. Hassani, N. Abdolhosein, F. Morteza, G. Ahangari, Mechanical properties of epoxy/basalt polymer concrete: Experimental and analytical study, *Struct. Concr.* 19 (2018) 366–373. https://doi.org/10.1002/suco.201700003.
[60] A. Kumar, G. Singh, N. Bala, Evaluation of flexural strength of epoxy polymer concrete with red mud and fly ash, *Int. J. Curr. Eng. Technol.* 13 (2013) 1799–1803.
[61] M.R. Mohamed, S.N. Aldeen, R. Abdulrazza, A study of compression strength and flexural strength for polymer concrete, *Iraqi J. Sci.* 57 (2016) 2677–2684.
[62] M.C.S. Ribeiro, C.M.L. Tavares, M. Figueiredo, A.J.M. Ferreira, A.A. Fernandes, Bending characteristics of resin concretes, *Mater. Res.* 6 (2003) 247–254. https://doi.org/10.1590/S1516-14392003000200021.
[63] A.J.M. Ferreira, A.T. Marques, M.C.S. Ribeiro, P.R. N, Flexural performance of polyester and epoxy polymer mortars under severe thermal conditions, *Cem. Concr. Compos.* 26 (2004) 803–809.
[64] H. Abdel-Fattah, M.M. El-Hawary, Flexural behavior of polymer concrete, *Constr. Build. Mater.* 13 (1999) 253–262. https://doi.org/10.1016/S0950-0618(99)00030-6.
[65] M.H. Niaki, A. Fereidoon, M.G. Ahangari, Effect of basalt, silica sand and fly ash on the mechanical properties of quaternary polymer concretes, *Bull. Mater. Sci.* 41 (2018) 69–80. https://doi.org/10.1007/s12034-018-1582-6.
[66] A.J.M. Ferreira, C. Tavares, C. Ribeiro, Flexural properties of polyester resin concretes, *J. Polym. Eng.* 20 (2000) 459–468. https://doi.org/10.1515/POLYENG.2000.20.6.459.
[67] M. Haidar, E. Ghorbel, H. Toutanji, Optimization of the formulation of micro-polymer concretes, *Constr. Build. Mater.* 25 (2011) 1632–1644. https://doi.org/http://dx.doi.org/10.1016/j.conbuildmat.2010.10.010.
[68] K. Kobayashi, T. Ito, Several physical properties of resin concrete, in: *1st Int. Congr. Polym. Concr. Concr.*, London, UK, 1975: pp. 236–240.
[69] J.B. Chem, J.J. Sokołowska, Technological properties of polymer concrete containing vinyl-ester resin waste mineral powder, *J. Build. Chem.* 1 (2016) 84–91.
[70] J.M.L. Reis, D.C. Moreira, L.C.S. Nunes, L.A. Sphaier, Experimental investigation of the mechanical properties of polymer mortars with nanoparticles, *Mater. Sci. Eng. A.* 528 (2011) 6083–6085. https://doi.org/10.1016/j.msea.2011.04.054.
[71] M.M. Shokrieh, M. Heidari-Rarani, M. Shakouri, E. Kashizadeh, Effects of thermal cycles on mechanical properties of an optimized polymer concrete, *Constr. Build. Mater.* 25 (2011) 3540–3549. https://doi.org/10.1016/j.conbuildmat.2011.03.047.
[72] J.M.L. Reis, Fracture and flexural characterization of natural fiber-reinforced polymer concrete, *Constr. Build. Mater.* 20 (2006) 673–678. https://doi.org/10.1016/j.conbuildmat.2005.02.008.
[73] O. Elalaoui, E. Ghorbel, V. Mignot, M. Ben Ouezdou, Mechanical and physical properties of epoxy polymer concrete after exposure to temperatures up to 250°C, *Constr. Build. Mater.* 27 (2012) 415–424. https://doi.org/10.1016/j.conbuildmat.2011.07.027.
[74] J.P. Gorninski, D.C. Dal Molin, C.S. Kazmierczak, Study of the modulus of elasticity of polymer concrete compounds and comparative assessment of polymer concrete and portland cement concrete, *Cem. Concr. Res.* 34 (2004) 2091–2095. https://doi.org/10.1016/j.cemconres.2004.03.012.
[75] C. Vipulanandan, Mechanical behaviour of polymer concrete systems, *Mater. Struct.* 21 (1988) 268–277.

[76] A. Sand, Effect of gravel-sand ratio on physico-mechanical, thermal and macro-structural properties of micro epoxy polymer concrete based on a mixture of alluvial-dune sand, *Open Civ. Eng. J.* 14 (2020) 247–261. https://doi.org/10.2174/1874149502014010247.
[77] S. Orak, Investigation of vibration damping on polymer concrete with polyester resin, *Cem. Concr. Res.* 30 (2000) 171–174. https://doi.org/10.1016/S0008-8846(99)00225-2.
[78] R.D. Maksimov, L.A. Jirgens, E.Z. Plume, J.O. Jansons, Water resistance of polyester polymer concrete, *Mech. Compos. Mater.* 39 (2003) 99–110. https://doi.org/10.1023/A:1023407910034.
[79] K.-Y. Yeon, Y.-S. Choi, S.-H. Hyun, Properties of recycled polymer concrete using crushed polymer concrete as an aggregate, in: *2nd Int. Conf. Sustain. Constr. Mater. Technol.*, Ancona, Italy, 2010: pp. 1299–1308.
[80] F. Carrión, L. Montalbán, J.I. Real, T. Real, Mechanical and physical properties of polyester polymer concrete using recycled aggregates from concrete sleepers, *Sci. World J.* 2014 (2014) 1–10. https://doi.org/http://dx.doi.org/10.1155/2014/526346.
[81] K. Yeon, Stress – strain curve modeling and length effect of polymer concrete subjected to flexural compressive stress, *J. Appl. Polym. Sci.* 114 (2009) 3819–3826. https://doi.org/10.1002/app.
[82] K. Rebeiz, S. Serhal, A. Craft, Properties of polymer concrete using fly ash, *J. Mater. Civ. Eng.* 16 (2004) 15–19. https://doi.org/10.1061/(ASCE)0899-1561(2004)16:1(15).
[83] B.-W. Jo, S.-K. Park, D.-K. Kim, Mechanical properties of nano-MMT reinforced polymer composite and polymer concrete, *Constr. Build. Mater.* 22 (2008) 14–20. https://doi.org/10.1016/j.conbuildmat.2007.02.009.
[84] R.D. Maksimov, L. Jirgens, J. Jansons, E. Plume, Mechanical properties of polyester polymer-concrete, *Mech. Compos. Mater.* 35 (1999) 99–110. https://doi.org/10.1007/BF02257239.
[85] W. Bai, J. Zhang, P. Yan, X. Wang, Study on vibration alleviating properties of glass fiber reinforced polymer concrete through orthogonal tests, *Mater. Des.* 30 (2009) 1417–1421. https://doi.org/https://doi.org/10.1016/j.matdes.2008.06.028.
[86] N. Ahn, D.K. Park, J. Lee, M.K. Lee, Structural test of precast polymer concrete, *J. Appl. Polym. Sci.* 114 (2009) 1370–1376. https://doi.org/10.1002/app.
[87] H. Haddad, *Optimisation of Polymer Concrete for the Manufacture of the Precision Tool Machines Bases*, Swinburne University of Technology, 2013.
[88] J. Wang, Q. Dai, S. Guo, R. Si, Mechanical and durability performance evaluation of crumb rubber-modified epoxy polymer concrete overlays, *Constr. Build. Mater.* 203 (2019) 469–480. https://doi.org/10.1016/j.conbuildmat.2019.01.085.
[89] G. Martínez-barrera, E.V. Santiago, S.H. López, O. Gencel, F. Ureña-nuñez, Polypropylene fibers as reinforcements of polyester-based composites, *Int. J. Polym. Sci.* 2013 (2013) 1–6. https://doi.org/10.1155/2013/143894.
[90] Y. Shen, J. Huang, X. Ma, F. Hao, J. Lv, Experimental study on the free shrinkage of lightweight polymer concrete incorporating waste rubber powder and ceramsite, *Compos. Struct.* 242 (2020) 112152. https://doi.org/10.1016/j.compstruct.2020.112152.
[91] Y. Shen, X. Ma, J. Huang, F. Hao, J. Lv, M. Shen, Near-zero restrained shrinkage polymer concrete incorporating ceramsite and waste rubber powder, *Cem. Concr. Compos.* 110 (2020) 103584. https://doi.org/10.1016/j.cemconcomp.2020.103584.
[92] H. Sanaei Ataabadi, A. Zare, H. Rahmani, A. Sedaghatdoost, E. Mirzaei, Lightweight dense polymer concrete exposed to chemical condition and various temperatures: An experimental investigation, *J. Build. Eng.* 34 (2021) 101878. https://doi.org/https://doi.org/10.1016/j.jobe.2020.101878.
[93] Susilawati, A.U. Husna, B. Ferdiansyah, Synthesis and characterization of polymer concrete withpumice aggregate and singkut leaf fiber as filler, *J. Phys. Conf. Ser.* (2021). https://doi.org/10.1088/1742-6596/1811/1/012041.

[94] J. Toma, Reinforced polymer concrete: Physical properties of the matrix and static/dynamic bond behaviour, 27 (2005) 934–944. https://doi.org/10.1016/j.cemconcomp.2005.06.004.
[95] P.S. Vijaya, B. Reddy, V. Santhosha, Experimental study on fibre reinforced polymer concrete, *Int. J. Appl. Eng. Res.* 13 (2018) 11844–11856.
[96] M.M.R. Taha, M. Genedy, Y. Ohama, Polymer concrete, in: *Dev. Formul. Reinf. Concr., Second Edi*, Woodhead Publishing, 2019: pp. 391–408. https://doi.org/10.1016/B978-0-08-102616-8.00017-4.
[97] C. Sung, Y. Kim, Void ratio and durability properties of porous polymer concrete using recycled aggregate with binder contents for permeability pavement, (2012). https://doi.org/10.1002/app.
[98] A.M. Hameed, M.T. Hamza, Characteristics of polymer concrete produced from wasted construction materials, *Energy Procedia.* 157 (2019) 43–50. https://doi.org/10.1016/j.egypro.2018.11.162.
[99] A. Heidari, T. Shams, M.R. Adlparvar, Mechanical properties of polymer concrete containing aluminum hydroxide using additives, *Sharif J. Civ. Eng.* 37(2) (2021) 127–134. https://doi.org/10.24200/j30.2021.55519.2743.
[100] I.S. Al-Haydari, G.G. Masood, S.A. Mohamad, H.M.N. Khudhur, Stress-strain behavior of sustainable polyester concrete with different types of recycled aggregate, Mater. Today Proc. 46 (2019) 5160–5166. https://doi.org/10.1016/j.matpr.2021.01.591.
[101] K.-S. Yeon, Polymer concrete as construction materials, *Int. J. Soc. Mater. Eng. Resour.* 17 (2010) 107–111.
[102] M. Rokbi, B. Baali, Z.E.A. Rahmouni, H. Latelli, Mechanical properties of polymer concrete made with jute fabric and waste marble powder at various woven orientations, *Int. J. Environ. Sci. Technol.* 16 (2019) 5087–5094. https://doi.org/10.1007/s13762-019-02367-7.
[103] E. Sab, R. Udroiu, P. Bere, I. Buransk, A novel polymer concrete composite with GFRP waste: Applications, morphology, and porosity characterization, *Appl. Sci.* 10 (2020) 2060. https://doi.org/10.3390/app10062060.
[104] P. De Pós-graduação, E. Mecânica, Mechanical characterization using optical fiber sensors of polyester polymer concrete made with recycled aggregates 3. *Experimental Procedure and Results*, 12 (2009) 269–271.
[105] J.M.L. dos Reis, M.A.G. Jurumenha, Experimental investigation on the effects of recycled aggregate on fracture behavior of polymer concrete, *Mater. Res.* 14 (2011) 326–330. https://doi.org/10.1590/S1516-14392011005000060.
[106] M. Saribiyik, A. Piskin, A. Saribiyik, The effects of waste glass powder usage on polymer concrete properties, *Constr. Build. Mater.* 47 (2013) 840–844. https://doi.org/https://doi.org/10.1016/j.conbuildmat.2013.05.023.
[107] P. Ogrodnik, Physico-mechanical properties and microstructure of polymer concrete with recycled glass aggregate, *Materials (Basel).* 11 (2018) 1–15. https://doi.org/10.3390/ma11071213.
[108] K. Jafari, V. Toufigh, Experimental and analytical evaluation of rubberized polymer concrete, *Constr. Build. Mater.* 155 (2017) 495–510. https://doi.org/10.1016/j.conbuildmat.2017.08.097.
[109] M. Martínez-lópez, G. Martínez-barrera, C. Barrera-díaz, F. Ureña-núñez, Waste Tetra Pak particles from beverage containers as reinforcements in polymer mortar : Effect of gamma irradiation as an interfacial coupling factor, *Constr. Build. Mater.* 121 (2016) 1–8. https://doi.org/10.1016/j.conbuildmat.2016.05.153.
[110] M. Martínez-lópez, G. Martínez-barrera, C. Barrera-díaz, F. Ureña-núñez, W. Brostow, Waste materials from Tetra pak packages as reinforcement of polymer concrete, *Int. J. Polym. Sci.* 2015 (2015) 1–8.

[111] G. Zanvettor, M. Barbuta, A. Rotaru, L. Bejan, Tensile properties of green polymer concrete, *Procedia Manuf.* 32 (2019) 248–252. https://doi.org/10.1016/j.promfg.2019.02.210.
[112] N. Kępczak, R. Rosik, M. Urbaniak, Material-removing machining wastes as a filler of a polymer concrete (industrial chips as a filler of a polymer concrete), *Sci. Eng. Compos. Mater.* 28 (2021) 343–351. https://doi.org/10.1515/secm-2021-0035.
[113] M. Bărbuţă, M. Harja, I. Baran, Comparison of mechanical properties for polymer concrete with different types of filler, *J. Mater. Civ. Eng.* 22 (2010) 696–701. https://doi.org/10.1061/(ASCE)MT.1943-5533.0000069.
[114] H.A. Bulut, R. Şahin, A study on mechanical properties of polymer concrete containing electronic plastic waste, *Compos. Struct.* 178 (2017) 50–62. https://doi.org/10.1016/j.compstruct.2017.06.058.
[115] P.J.R.O. Novoa, M.C.S. Ribeiro, A.J.M. Ferreira, A.T. Marques, Mechanical characterization of lightweight polymer mortar modified with cork granulates, *Compos. Sci. Technol.* 64 (2004) 2197–2199. https://doi.org/10.1016/j.compscitech.2004.03.006.
[116] A. Seco, A.M. Echeverría, S. Marcelino, B. García, S. Espuelas, Durability of polyester polymer concretes based on metallurgical wastes for the manufacture of construction and building products, *Constr. Build. Mater.* 240 (2020) 117907. https://doi.org/10.1016/j.conbuildmat.2019.117907.
[117] A. Seco, A. Maria, S. Marcelino, B. Garc, S. Espuelas, Characterization of fresh and cured properties of polymer concretes based on two metallurgical wastes, *Appl. Sci.* 10 (2020) 825. https://doi.org/10.3390/app10030825.
[118] Y. Ohama, Mix proportions and properties of polyester resin concretes, *Am. Concr. Inst.* 70 (1973) 283–294.
[119] V. Toufigh, D. Ph, S.M. Shirkhorshidi, M. Hosseinali, Experimental investigation and constitutive modeling of polymer concrete and sand interface, *Int. J. Geomech.* 17 (2016) 1–11. https://doi.org/10.1061/(ASCE)GM.1943-5622.0000695.
[120] M. Muthukumar, D. Mohan, M. Rajendran, Optimization of mix proportions of mineral aggregates using Box Behnken design of experiments, *Cem. Concr. Compos.* 25 (2003) 751–758. https://doi.org/10.1016/S0958-9465(02)00116-6.
[121] M. Golestaneh, G. Amini, G.D. Najafpour, M.A. Beygi, Evaluation of mechanical strength of epoxy polymer concrete with silica powder as filler, *World Appl. Sci. J.* 9 (2010) 216–220.
[122] G. Matinez Barrera, E. Vigueras Santiago, O. Gencel, H.E. Hagg Lobland, Polymer concretes: A description and methods for modification and improvement, *Mater. Educ.* 33 (2011) 37–52.
[123] V. Matikainen, K. Niemi, H. Koivuluoto, P. Vuoristo, Abrasion, erosion and cavitation erosion wear properties of thermally sprayed alumina based coatings, Coatings. 4 (2014) 18–36. https://doi.org/10.3390/coatings4010018.
[124] L. Xu, Y. Hu, H. Wu, J. Tian, J. Liu, Z. Gao, L. Wang, Surface crystal chemistry of spodumene with different size fractions and implications for flotation, *Sep. Purif. Technol.* 169 (2016) 33–42. https://doi.org/10.1016/j.seppur.2016.06.005.
[125] M.L. Hjuler, V.F. Hansen, I.D.A.L. Fabricius, Interpretational challenges related to studies of chalk particle surfaces in scanning and transmission electron microscopy, *Bull. Geol. Soc. Denmark.* 66 (2018) 151–166.
[126] B. Barra, L. Momm, Y. Guerrero, L. Bernucci, Characterization of granite and limestone powders for use as fillers in bituminous mastics dosage, *Ann. Brazilian Acad. Sci.* 86 (2014) 995–1002. https://doi.org/10.1590/0001-3765201420130165.
[127] K.E. Alyamaç, R. Ince, A preliminary concrete mix design for SCC with marble powders, *Constr. Build. Mater.* 23 (2009) 1201–1210. https://doi.org/https://doi.org/10.1016/j.conbuildmat.2008.08.012.

[128] J. Jiang, T. Feng, H. Chu, Y. Wu, F. Wang, W. Zhou, Z. Wang, Quasi-static and dynamic mechanical properties of eco-friendly ultra-high-performance concrete containing aeolian sand, *Cem. Concr. Compos.* 97 (2019) 369–378. https://doi.org/10.1016/j.cemconcomp.2019.01.011.

[129] S.T. Erdoğan, A.M. Forster, P.E. Stutzman, E.J. Garboczi, Particle-based characterization of Ottawa sand: Shape, size, mineralogy, and elastic moduli, *Cem. Concr. Compos.* 83 (2017) 36–44. https://doi.org/10.1016/j.cemconcomp.2017.07.003.

[130] M. Muthukumar, D. Mohan, Optimization of mechanical properties of polymer concrete and mix design recommendation based on design of experiments, *J. Appl. Polym. Sci.* 94 (2004) 1107–1116.

[131] R. Bedi, R. Chandra, S.P. Singh, Mechanical properties of polymer concrete, *J. Compos.* 2013 (2013) 1–12. https://doi.org/10.1155/2013/948745.

[132] H. Haddad, M. Al Kobaisi, Optimization of the polymer concrete used for manufacturing bases for precision tool machines, *Compos. Part B Eng.* 43 (2012) 3061–3068. https://doi.org/10.1016/j.compositesb.2012.05.003.

[133] J.P. Gorninski, D.C. Dal Molin, C.S. Kazmierczak, Strength degradation of polymer concrete in acidic environments, *Cem. Concr. Compos.* 29 (2007) 637–645. https://doi.org/10.1016/j.cemconcomp.2007.04.001.

[134] R.S. Raja, K. Manisekar, V. Manikandan, Effect of fly ash filler size on mechanical properties of polymer matrix composites, *Int. J. Mining, Metall. Mech. Eng.* 1 (2013) 34–37.

[135] B.-W. Jo, G.-H. Tae, C.-H. Kim, Uniaxial creep behavior and prediction of recycled-PET polymer concrete, *Constr. Build. Mater.* 21 (2007) 1552–1559. https://doi.org/http://dx.doi.org/10.1016/j.conbuildmat.2005.10.003.

[136] S. Park, B. Jo, J. Park, J. Choi, Fracture behaviour of polymer concrete reinforced with carbon and nylon fibres, *Adv. Cem. Res.* 22 (2010) 45–51. https://doi.org/10.1680/adcr.2008.22.1.45.

[137] J. Yeon, Deformability of bisphenol A-type epoxy resin-based polymer concrete with different hardeners and fillers, *Appl. Sci.* 10 (2020) 1336. https://doi.org/10.3390/app10041336.

[138] J. Mirza, N. Hafizah, A. Khalid, M. Warid, Effectiveness of palm oil fuel ash as micro-filler in polymer concrete, *J. Teknol. Full.* 16 (2015) 75–80.

[139] A. Garbacz, J.J. Sokołowska, Concrete-like polymer composites with fly ashes – Comparative study, *Constr. Build. Mater.* 38 (2013) 689–699. https://doi.org/10.1016/j.conbuildmat.2012.08.052.

[140] P. Mechanics, R. Engineering, Prediction of structural performance of vinyl ester polymer concrete using FEM elasto-plastic model, *Materials (Basel).* 13 (2020) 4034. https://doi.org/10.3390/ma13184034.

[141] H. Huang, H. Pang, J. Huang, H. Zhao, B. Liao, Synthesis and characterization of ground glass fiber reinforced polyurethane-based polymer concrete as a cementitious runway repair material, *Constr. Build. Mater.* 242 (2020) 117221. https://doi.org/10.1016/j.conbuildmat.2019.117221.

[142] M. Jamshidi, A.R. Pourkhorshidi, A comparative study on physical/mechanical properties of polymer concrete and portland cement concrete, *Asian J. Civ. Eng.* 11 (2010) 421–432.

[143] S. Ahn, E. Jeon, H. Koh, H. Kim, J. Park, Identification of stiffness distribution of fatigue loaded polymer concrete through vibration measurements, *Compos. Struct.* 136 (2016) 11–15. https://doi.org/10.1016/j.compstruct.2015.09.026.

[144] N.D. Shiri, S. Bhat, K.C. Babisha, K.M. Moger, M.P. D, C.J. Menezes, Taguchi analysis on the compressive strength behaviour of waste plastic-rubber composite materials, *Am. J. Mater. Sci.* 6 (2016) 88–93. https://doi.org/10.5923/c.materials.201601.17.

[145] B.W. Jo, C.H. Kim, J.H. Lim, Characteristics of cement mortar with nano-SiO 2 particles, *Constr. Build. Mater.* 21 (2007) 1351–1355. https://doi.org/10.1016/j.conbuildmat.2005.12.020.

[146] M.R.R. Hamester, P.S. Balzer, D. Becker, Characterization of calcium carbonate obtained from oyster and mussel shells and incorporation in polypropylene, *Mater. Res.* 15 (2012) 204–208. https://doi.org/10.1590/S1516-14392012005000014.

[147] M. Pettinato, D. Mukherjee, S. Andreoli, E.R. Minardi, V. Calabro, S. Curcio, S. Chakraborty, Industrial waste-an economical approach for adsorption of heavy metals from ground water, *Am. J. Eng. Appl. Sci.* 8 (2015) 48–56. https://doi.org/10.3844/ajeassp.2015.48.56.

[148] M.R. Karim, M.F.M. Zain, M. Jamil, F.C. Lai, Fabrication of a non-cement binder using slag, palm oil fuel ash and rice husk ash with sodium hydroxide, *Constr. Build. Mater.* 49 (2013) 894–902. https://doi.org/https://doi.org/10.1016/j.conbuildmat.2013.08.077.

[149] E.M. Van Der Merwe, L.C. Prinsloo, C.L. Mathebula, H.C. Swart, E. Coetsee, F.J. Doucet, Surface and bulk characterization of an ultrafine South African coal fly ash with reference to polymer applications, *Appl. Surf. Sci.* 317 (2014) 73–83. https://doi.org/10.1016/j.apsusc.2014.08.080.

[150] R. Siddique, M.I. Khan, Silica fume, in: *Suppl. Cem. Mater.*, Springer, Berlin, 2011: p. 288. https://doi.org/10.1007/978-3-642-17866-5.

[151] L. Black, Low clinker cement as a sustainable construction material, in: J.M.B.T.-S. of C.M. (Second) E. Khatib (Ed.), *Sustain. Constr. Mater.*, 2nd ed., Woodhead Publishing Series in Civil and Structural Engineering, Leeds, United Kingdom, 2016: pp. 415–457. https://doi.org/https://doi.org/10.1016/B978-0-08-100370-1.00017-2.

[152] I.Z. Yildirim, M. Prezzi, Chemical, mineralogical, and morphological properties of steel slag, *Adv. Civ. Eng.* 2011 (2011) 1–13. https://doi.org/10.1155/2011/463638.

[153] M. Tossavainen, F. Engstrom, Q. Yang, N. Menad, M. Lidstrom Larsson, B. Bjorkman, Characteristics of steel slag under different cooling conditions, *Waste Manage.* 27 (2007) 1335–1344. https://doi.org/10.1016/j.wasman.2006.08.002.

[154] E. Hwang, J. Kim, S. Park, Physical properties of polyester polymer concrete composite using RCSS fine aggregate, *Adv. Mater. Res.* 687 (2013) 219–228. https://doi.org/10.4028/www.scientific.net/AMR.687.219.

[155] H.M. Hamada, G.A. Jokhio, F.M. Yahaya, A.M. Humada, Y. Gul, The present state of the use of palm oil fuel ash (POFA) in concrete, *Constr. Build. Mater.* 175 (2018) 26–40. https://doi.org/https://doi.org/10.1016/j.conbuildmat.2018.03.227.

[156] S.A. Bernal, E.D. Rodríguez, A.P. Kirchheim, J.L. Provis, Management and valorisation of wastes through use in producing alkali-activated cement materials, *J. Chem. Technol. Biotechnol.* 91 (2016) 2365–2388. https://doi.org/10.1002/jctb.4927.

[157] J. Ambroise, M. Chabannet, J. Pe, Valorization of automotive shredder residue in building materials, 34 (2004) 557–562. https://doi.org/10.1016/j.cemconres.2003.09.004.

[158] R. Diaconescu, M. Barbuta, M. Harja, Prediction of properties of polymer concrete composite with tire rubber using neural networks, *Mater. Sci. Eng. B.* 178 (2013) 1259–1267. https://doi.org/10.1016/j.mseb.2013.01.014.

[159] I.B. Topçu, T. Uygunoglu, Sustainability of using waste rubber in concrete, in: J.M.B.T.-S. of C.M. (Second) E. Khatib (Ed.), *Sustain. Constr. Mater.*, 2nd ed., Woodhead Publishing, 2016: pp. 597–623. https://doi.org/https://doi.org/10.1016/B978-0-08-100370-1.00023-8.

[160] J.P. Gorninski, D.C. Dal Molin, C.S. Kazmierczak, Comparative assessment of isophtalic and orthophtalic polyester polymer concrete: Different costs, similar mechanical properties and durability, *Constr. Build. Mater.* 21 (2007) 546–555. https://doi.org/https://doi.org/10.1016/j.conbuildmat.2005.09.003.

[161] L. Czarnecki, *Polymer Concretes*, Arkady, Warszawa; [in Polish], 1982.

[162] M. Bărbuţă, M. Harja, Experimental study on the characteristics of polymer concrete with epoxy resin, *Bul. Inst. Politeh.* 54 (2008) 53–59.

[163] J.J. Sokołowska, Long-term investigation on the compressive strength of polymer concrete with fly ash, in: M.M.R. Taha (Ed.), *Int. Congr. Polym. Concr.*, Springer, Cham, Washington, DC, 2018: pp. 275–281. https://doi.org/https://doi.org/10.1007/978-3-319-78175-4_34.

[164] C. Atzeni, L. Massidda, U. Sanna, Mechanical properties of epoxy mortars with fly ash as filler, 12 (1990) 3–8.

[165] S.J. Julia, W.P. Paweł, Ł. Paweł, K. Kamila, Effect of perlite waste powder on chemical resistance of polymer concrete composites, *Adv. Mater. Res.* 1129 (2015) 516–522. https://doi.org/10.4028/www.scientific.net/AMR.1129.516.

[166] K. Shi-cong, P. Chi-sun, A novel polymer concrete made with recycled glass aggregates, fly ash and metakaolin, *Constr. Build. Mater.* 41 (2013) 146–151. https://doi.org/10.1016/j.conbuildmat.2012.11.083.

[167] K.S. Rebeiz, Precast use of polymer concrete using unsaturated polyester resin based on recycled PET waste, *Constr. Build. Mater.* 10 (1996) 215–220. https://doi.org/https://doi.org/10.1016/0950-0618(95)00088-7.

[168] T. Broniewski, Z. Jamrozy, J. Kapko, Long life strength polymer concrete, in: *1st Int. Congr. Polym. Concr.*, London, UK, 1975: pp. 179–184.

[169] J. Sim, C. Park, D.Y. Moon, Characteristics of basalt fiber as a strengthening material for concrete structures, *Compos. Part B Eng.* 36 (2005) 504–512. https://doi.org/10.1016/j.compositesb.2005.02.002.

[170] V. Fiore, T. Scalici, G. Di Bella, A. Valenza, A review on basalt fiber and its composites, *Compos. Part B Eng.* 74 (2015) 74–94. https://doi.org/10.1016/j.compositesb.2014.12.034.

[171] H. Kim, Enhancement of thermal and physical properties of epoxy composite reinforced with basalt fiber, *Fibers Polym.* 14 (2013) 1311–1316. https://doi.org/10.1007/s12221-013-1311-0.

[172] E. Quagliarini, F. Monni, S. Lenci, F. Bondioli, Tensile characterization of basalt fiber rods and ropes: A first contribution, *Constr. Build. Mater.* 34 (2012) 372–380. https://doi.org/10.1016/j.conbuildmat.2012.02.080.

[173] K. Singha, A short review on basalt fiber, *Int. J. Text. Sci.* 1 (2012) 19–28. https://doi.org/10.5923/j.textile.20120104.02.

[174] M.H. Niaki, A. Fereidoon, M.G. Ahangari, Experimental study on the mechanical and thermal properties of basalt fiber and nanoclay reinforced polymer concrete, *Compos. Struct.* 191 (2018) 231–238. https://doi.org/10.1016/j.compstruct.2018.02.063.

[175] V. Dhand, G. Mittal, K.Y. Rhee, D. Hui, A short review on basalt fiber reinforced polymer composites, *Compos. Part B.* 73 (2015) 166–180. https://doi.org/10.1016/j.compositesb.2014.12.011.

[176] D. Asprone, E. Cadoni, F. Iucolano, A. Prota, Analysis of the strain-rate behavior of a basalt fiber reinforced natural hydraulic mortar, *Cem. Concr. Compos.* 53 (2014) 52–58. https://doi.org/10.1016/j.cemconcomp.2014.06.009.

[177] P. Xu, W. Li, Y. Yu, J. Shen, Analysis of chatter and turning errors of basalt fiber reinforced polymer concrete lathe, *J. Phys. Conf. Ser.* 1748 (2021) 1–7. https://doi.org/10.1088/1742-6596/1748/6/062058.

[178] R.N. Ratu, *Development of Polypropylene Fiber as Concrete Reinforcing Fiber*, The University of British Columbia, 2016.

[179] J.M.L. Reis, Mechanical characterization of fiber reinforced polymer concrete, *Mater. Res.* 8 (2005) 357–360.

[180] Z. Zheng, D. Feldman, Synthetic fibre-reinforced concrete, *Prog. Polym. Sci.* 20 (1995) 185–210. https://doi.org/https://doi.org/10.1016/0079-6700(94)00030-6.

[181] G. Martı, L.F. Giraldo, B.L. Lo, Effects of g radiation on fiber-reinforced polymer concrete, *Polym. Compos.* 29 (2008) 1244–1251. https://doi.org/10.1002/pc.
[182] G. Martínez-Barrera, A.L. Martínez-Hernández, C. Velasco-Santos, M. Martínez-López, J. Ortiz-Espinoza, J.M.L. dos Reis, Polypropylene fibre reinforced polymer concrete: Effect of gamma irradiation, *Polym. Polym. Compos.* 22 (2014) 787–792. https://doi.org/10.1177/096739111402200905.
[183] I.V. Onuegbu, D.K. Kepkugile, Characterization of polymer concrete made of coconut/glass fibers and natural silica sand from Niger Delta, Nigeria, *Int. J. Sci. Eng. Res.* 5 (2015) 776–780.
[184] F. Vogt, A. Gypser, F. Kleiner, A. Osburg, polymer concrete for a modular construction system: Investigation of mechanical properties and bond behaviour by means of X-ray CT, in: M.M.R. Taha (Ed.), *Int. Congr. Polym. Concr. (ICPIC)*, Springer, Cham, Washington, DC, 2018: pp. 255–260. https://doi.org/https://doi.org/10.1007/978-3-319-78175-4_31.
[185] J.M.L. Reis, Sisal fiber polymer mortar composites: Introductory fracture mechanics approach, *Constr. Build. Mater.* 37 (2012) 177–180. https://doi.org/https://doi.org/10.1016/j.conbuildmat.2012.07.088.
[186] M. Barbuta, A. Timu, L. Bejan, R.D. Bucur, Mechanical properties of fly Ash polymer concrete with different fibers, *Mater. Plast.* 55 (2018) 405–409.
[187] V. Toufigh, V. Toufigh, H. Saadatmanesh, S. Ahmari, Strength evaluation and energy-dissipation behavior of fiber-reinforced polymer concrete, *Adv. Civ. Eng. Mater.* 2 (2013) 622–636. https://doi.org/https://doi.org/10.1520/ACEM20130074.
[188] J.M. Reis, A.J. Ferreira, Assessment of fracture properties of epoxy polymer concrete reinforced with short carbon and glass fibers, *Constr. Build. Mater.* 18 (2004) 523–528. https://doi.org/10.1016/j.conbuildmat.2004.04.010.
[189] C. Vipulanandan, S. Mebarkia, Aggregates, fibers and coupling agents in polymer concrete, in: *1st Mater. Eng. Conf.*, Denver, Colorado, 1990: pp. 785–794.
[190] X.U. Ping, Y. Ying-hua, Research on steel-fibber polymer concrete machine tool structure, *J. Coal Sci. Eng.* 14 (2008) 689–692.
[191] K. Sett, C. Vipulanandan, Properties of polyester polymer concrete with glass and carbon fibers, *Mater. J.* 101 (2004) 30. https://doi.org/10.14359/12985.
[192] M. Harja, M.B. Arbut, Properties of fiber reinforced polymer concrete, *Bull. Polytech. Inst. Jassy.* 4 (2008) 13–22.
[193] J.M.L. Reis, A.J.M. Ferreira, A contribution to the study of the fracture energy of polymer concrete and fibre reinforced polymer concrete, *Polym. Test.* 23 (2004) 437–440. https://doi.org/10.1016/j.polymertesting.2003.09.008.
[194] J.M.L. dos Reis, Effect of textile waste on the mechanical properties of polymer concrete, *Mater. Res.* 12 (2009) 63–67.
[195] Y. Jin, J. Liu, X. Xie, L. Zhai, Different effects of steel fiber and nano-SiO2 on epoxy resin concrete, *IOP Conf. Ser. Earth Environ. Sci.* 186 (2018) 12040.
[196] G. Martínez-barrera, J. José, E. Martínez-cruz, M. Martínez-lópez, M.C.S. Ribeiro, C. Velasco-santos, H.E. Hagg, W. Brostow, Modified recycled tire fibers by gamma radiation and their use on the improvement of polymer concrete, *Constr. Build. Mater.* 204 (2019) 327–334. https://doi.org/10.1016/j.conbuildmat.2019.01.177.
[197] O. Petruška, J. Zajac, V. Molnár, G. Fedorko, J. Tkácˇ, The effect of the carbon fiber content on the flexural strength of polymer concrete testing samples and the comparison of polymer concrete and U-shaped steel profile damping, *Materials (Basel).* 12 (2019) 1917. https://doi.org/10.3390/ma12121917.
[198] N. Gupta, T.C. Lin, M. Shapiro, Clay-epoxy nanocomposites : Processing and properties, *J. Miner. Met. Mater. Soc.* 59 (2007) 61–65.

[199] Y. Rostamiyan, A. Fereidoon, A.H. Mashhadzadeh, Investigation of damping and toughness properties of epoxy-based nanocomposite using different reinforcement mechanisms: Polymeric alloying, nanofiber, nanolayered, and nanoparticulate materials, *Sci. Eng. Compos. Mater.* 22 (2015) 591–598. https://doi.org/10.1515/secm-2013-0305.
[200] A. Douba, M. Genedy, E. Matteo, U.F. Kandil, M.M.R. Taha, The significance of nanoparticles on bond strength of polymer concrete to steel, *Int. J. Adhes. Adhes.* 74 (2017) 77–85. https://doi.org/10.1016/j.ijadhadh.2017.01.001.
[201] A.J.M.F.M.C.S. Riberio, C.M.C. Pereira, S.P.B. Sousa, P.R.O. Novoa, Fire reaction and mechanical performance analyses of polymer concrete materials modified with micro and nano alumina particles, *Restor. Build. Monum.* 19 (2013) 195–202.
[202] A. Douba, *Mechanical Characterization of Polymer Concrete with Nanomaterials*, The University of New Mexico, Thesis for Master of Science, 2017.
[203] J.M.L. Reis, D.C. Moreira, L.C.S. Nunes, L.A. Sphaier, Evaluation of the fracture properties of polymer mortars reinforced with nanoparticles, *Compos. Struct.* 93 (2011) 3002–3005. https://doi.org/10.1016/j.compstruct.2011.05.002.
[204] S.M. Daghash, *New Generation Polymer Concrete Incorporating Carbon Nanotubes*, The University of New Mexico, Thesis for Master of Science, 2013.
[205] M.M.R. Taha, Nano-modified polymer concrete: A new material for smart and resilient structures, in: *Int. Congr. Polym. Concr.*, Springer, Cham, Washington, DC, 2018: pp. 61–73. https://doi.org/https://doi.org/10.1007/978-3-319-78175-4_6.
[206] A. Douba, M.R. Taha, PC with superior ductility using mixture of pristine and functionalized carbon nanotubes, in: M.M.R. Taha (Ed.), *Int. Congr. Polym. Concr. (ICPIC)*, Springer, Cham, 2018: pp. 291–297. https://doi.org/https://doi.org/10.1007/978-3-319-78175-4_36.
[207] M.M. Shokrieh, A.R. Kefayati, M. Chitsazzadeh, Fabrication and mechanical properties of clay/epoxy nanocomposite and its polymer concrete, *Mater. Des.* 40 (2012) 443–452. https://doi.org/10.1016/j.matdes.2012.03.008.
[208] O. Ben David, L. Banks-Sills, J. Aboudi, V. Fourman, R. Eliasi, T. Simhi, A. Shlayer, O. Raz, Evaluation of the mechanical properties of PMMA reinforced with carbon nanotubes – experiments and modeling, *Exp. Mech.* 54 (2014) 175–186. https://doi.org/10.1007/s11340-013-9792-8.
[209] J. Williams, W. Broughton, T. Koukoulas, S.S. Rahatekar, Plasma treatment as a method for functionalising and improving dispersion of carbon nanotubes in epoxy resins, *J. Mater. Sci.* 48 (2013) 1005–1013. https://doi.org/10.1007/s10853-012-6830-3.
[210] M. Theodore, M. Hosur, J. Thomas, S. Jeelani, Influence of functionalization on properties of MWCNT–epoxy nanocomposites, *Mater. Sci. Eng. A.* 528 (2011) 1192–1200. https://doi.org/https://doi.org/10.1016/j.msea.2010.09.095.
[211] Z. He, X. Zhang, M. Chen, M. Li, Y. Gu, Z. Zhang, Q. Li, Effect of the filler structure of carbon nanomaterials on the electrical, thermal, and rheological properties of epoxy composites, *J. Appl. Polym. Sci.* 129 (2013) 3366–3372. https://doi.org/10.1002/app.39096.
[212] N.G. Shimpi, R.U. Kakade, S.S. Sonawane, A.D. Mali, S. Mishra, Influence of nano-inorganic particles on properties of epoxy nanocomposites, *Polym. - Plast. Technol. Eng.* 50 (2011) 758–761. https://doi.org/10.1080/03602559.2010.551437.
[213] Y.K. Hamidi, L. Aktas, M.C. Altan, Effect of nanoclay content on void morphology in resin transfer molded composites, *J. Thermoplast. Compos. Mater.* 21 (2008) 141–163. https://doi.org/10.1177/0892705707083635.
[214] B.M. Tyson, R.K. Abu Al-Rub, A. Yazdanbakhsh, Z. Grasley, A quantitative method for analyzing the dispersion and agglomeration of nano-particles in composite materials, *Compos. Part B Eng.* 42 (2011) 1395–1403. https://doi.org/https://doi.org/10.1016/j.compositesb.2011.05.020.

[215] S. Chuah, Z. Pan, J.G. Sanjayan, C.M. Wang, W.H. Duan, Nano reinforced cement and concrete composites and new perspective from graphene oxide, *Constr. Build. Mater.* 73 (2014) 113–124. https://doi.org/10.1016/j.conbuildmat.2014.09.040.
[216] M. Emiroglu, A.E. Douba, R.A. Tarefder, U.F. Kandil, M.R. Taha, New polymer concrete with superior ductility and fracture toughness using alumina nanoparticles, *J. Mater. Civ. Eng.* 29 (2017) 1–9. https://doi.org/10.1061/(ASCE)MT.1943-5533.0001894.
[217] A. Douba, M. Emiroglu, U.F. Kandil, M.M.R. Taha, Very ductile polymer concrete using carbon nanotubes, *Constr. Build. Mater.* 196 (2019) 468–477. https://doi.org/10.1016/j.conbuildmat.2018.11.021.

3 Polymer Concrete Preparation

Abstract

Finding the appropriate and optimal weight percentage of materials in polymer concrete (PC) composition has always been the focus of researchers. To date, various methods have been proposed to find the optimal weight percentage of materials in PC. Furthermore, after selecting the appropriate materials and finding their weight contents, knowing the fabrication methods and related parameters will have a significant impact on the properties of the PC. This chapter attempts to discuss PC preparation effective factors such as materials preparation, design of PC mixture, fabrication strategy and important parameters, and also curing and post-curing effects. The detrimental effect of moisture and dust on the quality and properties of PC will also be discussed.

3.1 DESIGN OF MIXTURE

It is important to design an optimum combination for obtaining the polymer concrete (PC) with maximum strength. Because of incorporating various fillers and aggregates in PC mixture, some dependent and independent variables like polymer, micro filler, aggregate and fiber weight content, aggregates, and other fillers size, play an important role in PC characteristics.

Design of experiments (DOE) are statistical approaches that can be used for optimizing multi-variable systems or if more than one response is of importance. Authors implemented some design strategies for minimizing the number of experiments and or optimizing a specific property, such as the Taguchi approach [1–4], mixture design approach [5,6], response surface method, and Box Behnken DOE [7–9].

Shokrieh et al. [10] considered an aggregate dimension, resin, and fibers wt% as design variables and simultaneously optimized the experimental mechanical properties according to the Taguchi method. By the use of the mixture design approach as a DOE technique for optimization of three variables (furan resin, optimized aggregate mix proportion, and silica powder), the experimental results were investigated. After determining the best mixture design for each specification and by combining the optimized results, a single combination with optimum value in all properties is recommended [6].

An optimum PC mixture should provide minimum void volume, which leads to denser packing of aggregates, maximum bulk density, and better properties. Also, the dense packing system requires less resin and reduces the cost of concrete production [7]. The authors applied the minimum void or maximum packing density method using the mix design on PC according to evaluating the air voids in aggregates. Require parameters of the process such as bulk specific gravity, saturated surface dry (SSD) specific gravity, apparent specific gravity, and absorption of fine aggregate were determined using ASTM C128 standard test method. The void content in the

DOI: 10.1201/9781003326311-3

sand was determined using specific gravity. The apparent density of sand is determined by the apparent specific gravity and the density of water. In the next step, the volume of pure material without voids and the air voids between sand particles were obtained. Finally, the attained air voids should be filled with various combinations of micro filler and resin to minimize the voids [11].

In literature [12], various contents of fine (0 and 4.75 mm) and coarse aggregates (4.75–19 mm) were mixed for achieving the highest dry bulk density based on ASTM C29. For this purpose, 30% fine aggregate and 70% coarse aggregate were mixed, and coarse aggregate weight fraction was fixed, whereas the fine aggregate was enhanced by 10 wt% in each stage. Then, the mixture was poured into a mold and was compacted using the rodding process. After weighing and calculating density, the mixture was poured back into a larger bucket, and 10 wt% fine aggregates were added to the mixture. This process was continued until a combination of 70% fine aggregate and 30% coarse aggregate was obtained. The combination of 50% of fine aggregates (0 and 4.75 mm) and 50% of coarse aggregates (4.75–19 mm) yielded the maximum dry bulk density and was applied as an appropriate aggregate mixture [12].

3.2 FABRICATION STRATEGIES

The PC preparation process depends on the type of material. We present a comprehensive flow chart of the PC preparation process containing nanoparticles, micro filler, fibers, and aggregates, as shown in Figure 3.1. Depending on the removal of any material, the corresponding stage will be eliminated. At all stages, materials should be prepared and mixed in specific temperatures, time, and relative humidity, depending on their specifications.

From Figure 3.1, nanomaterials, micro fillers, chopped fibers, and aggregates in order of priority, from smaller to larger particles, are added to the resin. Observing this helps to the homogeneous distribution of aggregates in concrete. In the case of nano-reinforced PC, nanoparticles are added to resin at first, using mentioned approaches. Then micro fillers, fibers, and aggregates can be dispersed in a mixture by using different automatic mixers such as a crosshead two-propelled mechanical

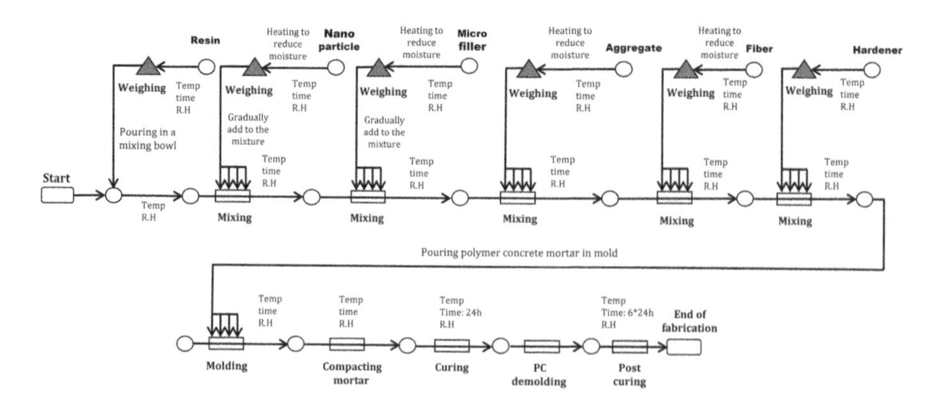

FIGURE 3.1 Polymer concrete preparation process flow chart.

mixer [13] and planetary type mortar mixer [14]. Because of incorporating low binder content in PC, mixing technology is a vital factor. Three mixing strategies for a specific composition were investigated in the literature [15]. The first one comprises of mixing 75% of the resin with the fine aggregate, and afterward the medium-sized aggregate, and finally, the coarse aggregate was introduced to the main mortar with the rest of the resin. In the second one, 20% of the resin was blended with coarse aggregate. Afterward, 30% of the resin was mixed with the micro filler separately. The medium-sized aggregate was mixed with the rest of the resin separately. In the last step, the three mixtures were blended. In the third mixing strategy, aggregates with different sizes were mixed and then added and mixed with the resin. It was observed that the first method results in maximum flexural strength.

The molding strategy has a considerable effect on the curing behavior and performance of PC. As it is mentioned, too many air bubbles and microvoids in the mixture reduce the performance of the PC. There are some methods for reducing the microvoids and emerging out the air bubbles. During the mixing process and even a few minutes after mixing, the tank should be vibrated by a vibration workbench. As a result, air bubbles emerged out of the concrete composite due to the vibration process and the remaining composite became more close-grained [16,17]. For determining the optimum frequency of the vibration table during the packing operation, various frequencies were implemented. The optimum vibration frequency for fabricating a PC with the maximum compressive strength was determined to be 18.9375 Hz [15]. Also, other pouring methods for reducing the air voids are summarized as follows [18]:

- **Compression molding technique:** Applying the pressure on PC mold helps the discharge of bubbles. Compression force can also be supplied using a metallic hammer [19] and the rodding method [12].
- **Negative pressure pouring technique:** Reducing external pressure helps the emerging out of bubbles. The vacuum pouring method is based on this technique.
- **Multi-layer pouring technique:** Decreasing the height of the resin upper bubble can effectively discharge the bubbles.
- **Add defoamers:** Reducing surface tension of the resin by adding defoaming agent helps the discharge of bubbles.

TABLE 3.1
Influence of implemented pressure on casting on the compressive strength of the PC [20]

Applied casting pressure (kg/cm^2)	Compressive strength (MPa)
Ordinary casting	18
300	48
500	50
700	58
800	60
900	55

The effect of various casting pressure on the compressive strength of the PC was determined in Ref. [20]. As presented in Table 3.1, applying pressure is accompanied by a significant enhancement in the compressive strength. It was due to the ejection of pores from the mixture and hence more rigid and compacted specimens due to the applied casting pressure. Also, a combination of the vibration method and applying a pressure of 10 kPa at the top of the specimen is recommended [21].

3.3 CURING EFFECT

Setting time (start with adding hardener) and needed time for obtaining the highest strength differed from a few minutes up to several hours. PC working life and curing time generally are controlled by the amount of hardening agent and casting temperature [22]. By increasing the curing age of PC, the cross-linking density of the matrix enhances, and the chain mobility of the PC decreases. Then the strength and the stiffness of the PC increases. Working life can be determined using Injection and Touching methods (JIS A 1186) [22].

It is mentioned that rapid curing is one of the main superiority of PC. PCs achieve 70% of their maximum strength after one day of curing at room temperature. They also achieve more than 85% of their seven days' average strength at only three days of curing, which is useful in precast applications [23]. As a result of the experimental study, the highest compressive, flexural, and tensile strengths were obtained after two, three, and seven days of curing [19]. Figure 3.2 demonstrates the effect of curing time on the compressive strength behavior of a typical PC. As can be observed, a large amount of ultimate compressive strength is achieved in the first 24 hours of curing. The increase in compressive strength will continue until the 28th day of curing [24].

Curing temperature and methods play a vital role in the properties of PC. Room temperature curing is an inexpensive and desirable curing method and is suitable for field uses. In this method, PC samples are detached from the mold after one day and

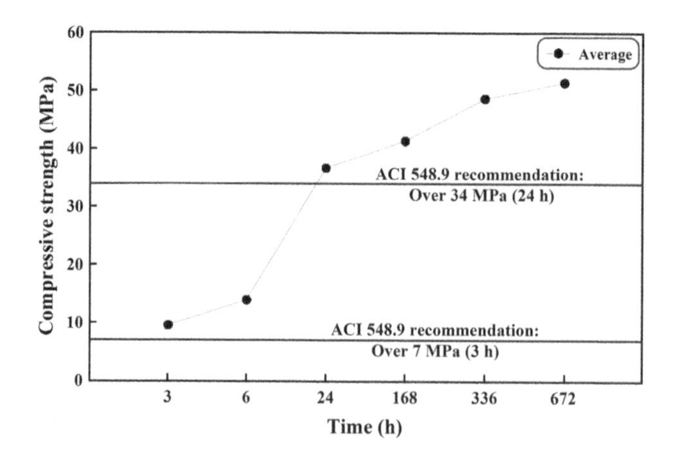

FIGURE 3.2 Variation of compressive strength of polysulfide PC with the curing time [24].

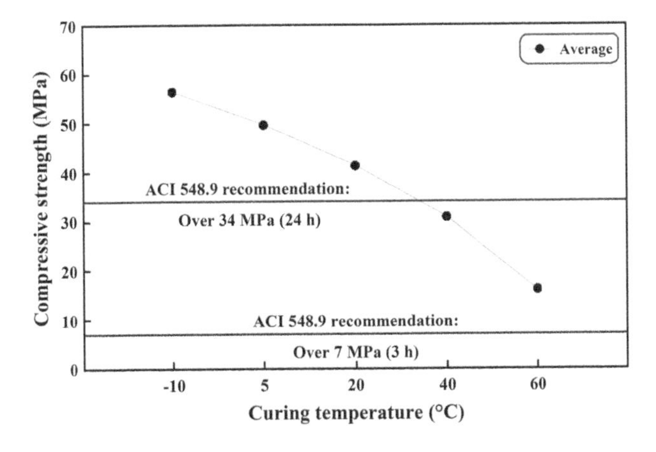

FIGURE 3.3 Variation of compressive strength of polysulfide PC with the curing temperature [24].

post-cured for six days. So concrete achieves the maximum performance after seven days of curing [5,13,25,26].

Effects of different curing temperatures on PC performance were investigated in the literature. Evaluating the effect of various curing temperatures on polysulfide PC proves that the cured specimen at −20°C has higher compressive and flexural strengths than the specimens cured at 5°C, 20°C, 40°C, and 60°C. Figure 3.3 presents the compressive strength behavior of this PC with various curing temperatures. Curing temperatures below zero degrees, by reducing the rate of chemical reactions, have a negative influence on the initial compressive and flexural strength of the samples. But the final strength obtained is greater [24].

By raising the curing temperature from 10°C to 30°C, the ultimate setting shrinkage of acrylic PC tended to enhance [27]. Decreasing the curing temperature from 20°C to −20°C in UP-MMA PC caused a sharp decline of strengths by 32.8% on average [23]. MMA-based PC can reach its full compressive strength at a low curing temperature of −20°C [28].

Because of applying thermoset resin in PCs, elevated temperature can accelerate and complete the curing process. Accelerated curing increased polymeric chain mobility and hence the rate of polymerization [29]. Curing in elevated temperatures implemented as a post-curing method in literature [30–34]. Also, gamma irradiation at different doses was implemented as an interfacial coupling factor after polymerization of the unsaturated polyester resin concrete, which improved the ductility of the PC [35].

3.4 MOISTURE AND DUST EFFECTS

Resin/aggregate adhesive bonding has an important effect on the performance of PC. Aggregates must be dehydrated and dustless for obtaining the best aggregate-resin

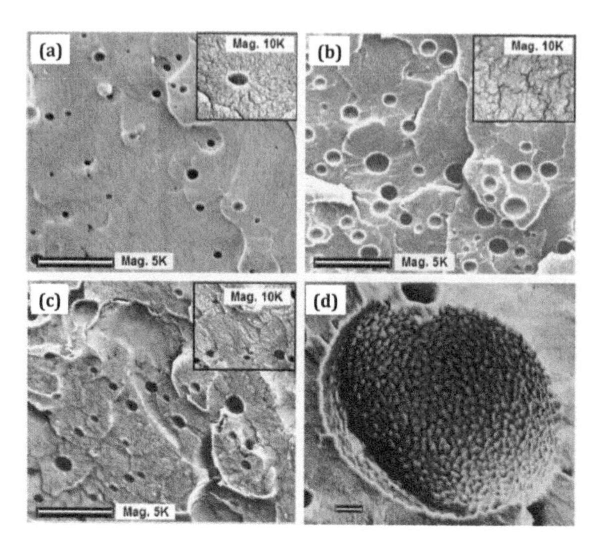

FIGURE 3.4 SEM micrographs of UPE resin including (a) 1%, (b) 3%, (c) 5% humidity (scale bar is 10 μm), and (d) the morphology of the voids left by water (scale bar is 1 μm) [36].

bonding. Thus, before adding aggregate within resin, they should be washed with water to remove dust and then dried in oven. The presence of moisture in any of the components of PC, including fillers and aggregates, even to a very small extent, will harm the properties of concrete.

SEM micrographs from fracture surface of unsaturated polyester resin containing three different moisture contents as variables reveal the destructive effect of moisture in the cured polymer (see Figure 3.4). As illustrated in Figure 3.4, by increasing the water content, some spherical voids were left by water droplets, which were unmixable in the resin increase. The elevated temperature during the resin curing process causes the humidity to evaporate and generate bubbles, which dramatically reduces the mechanical strength of the polymer [36]. So it is vital to eliminate humidity as possible.

If additives such as aggregates, micro fillers, or nanopowders absorb moisture, the bond strength between the resin and additives weakens and phase separation occurs due to hydrophobic polymer binder. Thus strength reduction of PC occurs. It is vital to keep the moisture content of all aggregates and fillers below 0.5 wt% before mixing [28,37].

The influence of humidity amount on thermal and mechanical characteristics and curing behavior of PC was evaluated by adding water to the aggregates. Results indicate that the influence of humidity in the micro filler and aggregate is the same as the influence on the resin. The added water is mainly presented on the hydrophilic surface of the aggregates, influencing the bond strength with the polymer matrix. Flexural and compressive strengths of the studied PC dramatically decrease by 38% and 55% with adding 5 wt% water, respectively (as shown in Figure 3.5a and 3.5b). But the coefficient of thermal expansions (CTEs) showed an increasing manner. Because of the chemical reaction of water molecules and the resin during the curing

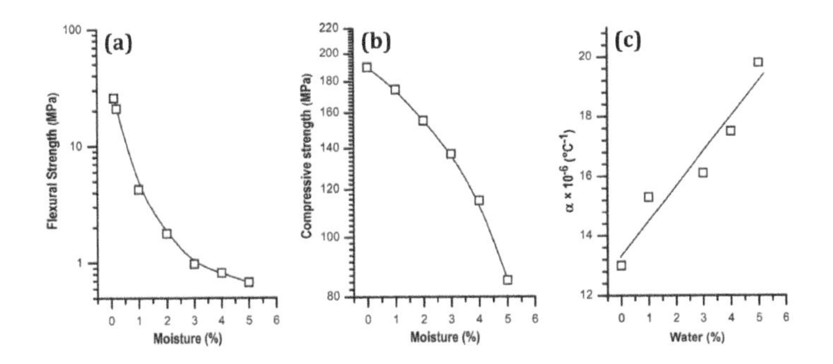

FIGURE 3.5 The effects of water contents on properties of UPE PC [36].

process, which influences the chemical properties of resin and reduces polymer/aggregate adhesion bonding, CTE of resin and PC increases with rising the moisture content (see Figure 3.5c), however, the strengths decrease. For achieving maximum strength, 0.002 wt% moisture in aggregates may be an acceptable level [36].

Mechanical properties, curing time, and workability of different PC made with wet and dry aggregates were investigated in Ref. [38]. Results proved that wet aggregates increased the workability of PC but decreased the flexural strength of epoxy PCs. The addition of metallic monomer powders: zinc diacrylate (ZDA) and calcium diacrylate (CDA) to MMA and polyester PCs improved the moisture resistance very significantly. Above-mentioned monomers may permit the possibility of using wet aggregates in PC construction [39].

REFERENCES

[1] K. Jafari, V. Toufigh, Experimental and analytical evaluation of rubberized polymer concrete, *Constr. Build. Mater.* 155 (2017) 495–510. https://doi.org/10.1016/j.conbuildmat.2017.08.097.

[2] A.J.M. Ferreira, C. Tavares, C. Ribeiro, Flexural properties of polyester resin concretes, *J. Polym. Eng.* 20 (2000) 459–468. https://doi.org/10.1515/POLYENG.2000.20.6.459.

[3] K. Jafari, M. Tabatabaeian, A. Joshaghani, T. Ozbakkaloglu, Optimizing the mixture design of polymer concrete: An experimental investigation, *Constr. Build. Mater.* 167 (2018) 185–187. https://doi.org/10.1016/j.conbuildmat.2018.01.191

[4] M. Heidari-Rarani, M.R.M. Aliha, M.M. Shokrieh, M.R. Ayatollahi, Mechanical durability of an optimized polymer concrete under various thermal cyclic loadings – An experimental study, *Constr. Build. Mater.* 64 (2014) 308–315. https://doi.org/10.1016/j.conbuildmat.2014.04.031.

[5] M.H. Niaki, A. Fereidoon, M.G. Ahangari, Effect of basalt, silica sand and fly ash on the mechanical properties of quaternary polymer concretes, *Bull. Mater. Sci.* 41 (2018) 69–80. https://doi.org/10.1007/s12034-018-1582-6.

[6] M. Muthukumar, D. Mohan, Optimization of mechanical properties of polymer concrete and mix design recommendation based on design of experiments, *J. Appl. Polym. Sci.* 94 (2004) 1107–1116.

[7] M. Muthukumar, D. Mohan, Studies on polymer concretes based on optimized aggregate mix proportion, *Eur. Polym. J.* 40 (2004) 2167–2177. https://doi.org/10.1016/j.eurpolymj.2004.05.004.

[8] M. Barbuta, D. Lepadatu, Mechanical characteristics investigation of polymer concrete using mixture design of experiments and response surface method, *J. Appl. Sci.* 8 (2008) 2242–2249.

[9] M. Muthukumar, D. Mohan, M. Rajendran, Optimization of mix proportions of mineral aggregates using Box Behnken design of experiments, *Cem. Concr. Compos.* 25 (2003) 751–758. https://doi.org/10.1016/S0958-9465(02)00116-6.

[10] M.M. Shokrieh, M. Heidari-Rarani, M. Shakouri, E. Kashizadeh, Effects of thermal cycles on mechanical properties of an optimized polymer concrete, *Constr. Build. Mater.* 25 (2011) 3540–3549. https://doi.org/10.1016/j.conbuildmat.2011.03.047.

[11] W. Lokuge, T. Aravinthan, Effect of fly ash on the behaviour of polymer concrete with different types of resin, *Mater. Des.* 51 (2013) 175–181. https://doi.org/10.1016/j.matdes.2013.03.078.

[12] V. Toufigh, M. Hosseinali, S. Masoud, Experimental study and constitutive modeling of polymer concrete's behavior in compression, *Constr. Build. Mater.* 112 (2016) 183–190. https://doi.org/10.1016/j.conbuildmat.2016.02.100.

[13] M.H. Niaki, A. Fereidoon, M.G. Ahangari, Experimental study on the mechanical and thermal properties of basalt fiber and nanoclay reinforced polymer concrete, *Compos. Struct.* 191 (2018) 231–238. https://doi.org/10.1016/j.compstruct.2018.02.063.

[14] A.S. Benosman, M. Mouli, H. Taibi, M. Belbachir, Mineralogical study of polymer-mortar composites with PET polymer by means of spectroscopic analyses, *Mater. Sci. Appl.* 3 (2012) 139–150. https://doi.org/10.4236/msa.2012.33022.

[15] H. Haddad, I. Sbarski, Optimization of moulding technology of polymer concrete used for manufacture precision tool machine bases, *J. Mater. Sci. Eng.* 7 (2018) 1–6. https://doi.org/10.4172/2169-0022.1000427.

[16] W. Bai, J. Zhang, P. Yan, X. Wang, Study on vibration alleviating properties of glass fiber reinforced polymer concrete through orthogonal tests, *Mater. Des.* 30 (2009) 1417–1421. https://doi.org/https://doi.org/10.1016/j.matdes.2008.06.028.

[17] I.V. Onuegbu, S.A. Amadi, The impact of crude oil contaminated natural sand on polymer concrete, *Curr. Stud. Comp. Educ. Sci. Technol.* 3 (2017) 217–226.

[18] J. Yin, J. Zhang, T. Wang, Y. Zhang, W. Wang, Experimental investigation on air void and compressive strength optimization of resin mineral composite for precision machine tool, *Polym. Compos.* 39 (2016) 457–466. https://doi.org/10.1002/pc.

[19] M. Haidar, E. Ghorbel, H. Toutanji, Optimization of the formulation of micro-polymer concretes, *Constr. Build. Mater.* 25 (2011) 1632–1644. https://doi.org/http://dx.doi.org/10.1016/j.conbuildmat.2010.10.010.

[20] M.E. Tawfik, S.B. Eskander, Polymer concrete from marble wastes and recycled poly(ethylene terephthalate), *J. Elastomers Plast.* 38 (2006) 65–79. https://doi.org/10.1177/0095244306055569.

[21] O. Elalaoui, E. Ghorbel, V. Mignot, M. Ben Ouezdou, Mechanical and physical properties of epoxy polymer concrete after exposure to temperatures up to 250°C, *Constr. Build. Mater.* 27 (2012) 415–424. https://doi.org/10.1016/j.conbuildmat.2011.07.027.

[22] N. Ahn, D.K. Park, J. Lee, M.K. Lee, Structural test of precast polymer concrete, *J. Appl. Polym. Sci.* 114 (2009) 1370–1376. https://doi.org/10.1002/app.

[23] S.-H. Hyun, J.H. Yeon, Strength development characteristics of UP-MMA based polymer concrete with different curing temperature, *Constr. Build. Mater.* 37 (2012) 387–397. https://doi.org/https://doi.org/10.1016/j.conbuildmat.2012.07.094.

[24] S. Hong, Influence of curing conditions on the strength properties of polysulfide polymer concrete, *Appl. Sci.* 7 (2017) 833. https://doi.org/10.3390/app7080833.

[25] W. Ferdous, A. Manalo, T. Aravinthan, G. Van Erp, Design of epoxy resin based polymer concrete matrix for composite railway sleeper, in: *23rd Australas. Conf. Mech. Struct. Mater.*, Southern Cross University ePublications, Byron Bay, Australia, 2014: pp. 137–142.

[26] M. Hassani, N. Abdolhosein, F. Morteza, G. Ahangari, Mechanical properties of epoxy/basalt polymer concrete: Experimental and analytical study, *Struct. Concr.* 19 (2018) 366–373. https://doi.org/10.1002/suco.201700003.

[27] K.-S. Yeon, J.H. Yeon, Y.-S. Choi, S.-H. Min, Deformation behavior of acrylic polymer concrete: Effects of methacrylic acid and curing temperature, *Constr. Build. Mater.* 63 (2014) 125–131. https://doi.org/https://doi.org/10.1016/j.conbuildmat.2014.04.051.

[28] S.-W. Son, J.H. Yeon, Mechanical properties of acrylic polymer concrete containing methacrylic acid as an additive, *Constr. Build. Mater.* 37 (2012) 669–679. https://doi.org/https://doi.org/10.1016/j.conbuildmat.2012.07.093.

[29] A. Douba, *Mechanical Characterization of Polymer Concrete with Nanomaterials*, The University of New Mexico, Thesis for Master of Science, 2017.

[30] P.J.R.O. Novoa, M.C.S. Ribeiro, A.J.M. Ferreira, A.T. Marques, Mechanical characterization of lightweight polymer mortar modified with cork granulates, *Compos. Sci. Technol.* 64 (2004) 2197–2199. https://doi.org/10.1016/j.compscitech.2004.03.006.

[31] M.C.S. Ribeiro, J.M.L. Reis, A.J.M. Ferreira, A.T. Marques, Thermal expansion of epoxy and polyester polymer mortars—Plain mortars and fibre-reinforced mortars, *Polym. Test.* 22 (2003) 849–857. https://doi.org/10.1016/S0142-9418(03)00021-7.

[32] M.M. Shokrieh, A.R. Kefayati, M. Chitsazzadeh, Fabrication and mechanical properties of clay/epoxy nanocomposite and its polymer concrete, *Mater. Des.* 40 (2012) 443–452. https://doi.org/10.1016/j.matdes.2012.03.008.

[33] J.M.L. Reis, Fracture and flexural characterization of natural fiber-reinforced polymer concrete, *Constr. Build. Mater.* 20 (2006) 673–678. https://doi.org/10.1016/j.conbuildmat.2005.02.008.

[34] J.M.L. Reis, D.C. Moreira, L.C.S. Nunes, L.A. Sphaier, Evaluation of the fracture properties of polymer mortars reinforced with nanoparticles, *Compos. Struct.* 93 (2011) 3002–3005. https://doi.org/10.1016/j.compstruct.2011.05.002.

[35] M. Martínez-lópez, G. Martínez-barrera, C. Barrera-díaz, F. Ureña-núñez, Waste Tetra Pak particles from beverage containers as reinforcements in polymer mortar: Effect of gamma irradiation as an interfacial coupling factor, *Constr. Build. Mater.* 121 (2016) 1–8. https://doi.org/10.1016/j.conbuildmat.2016.05.153.

[36] H. Haddad, M. Al, Influence of moisture content on the thermal and mechanical properties and curing behavior of polymeric matrix and polymer concrete composite, *Mater. Des.* 49 (2013) 850–856. https://doi.org/10.1016/j.matdes.2013.01.075.

[37] A.J.M.F.M.C.S. Riberio, C.M.C. Pereira, S.P.B. Sousa, P.R.O. Novoa, Fire reaction and mechanical performance analyses of polymer concrete materials modified with micro and nano alumina particles, *Restor. Build. Monum.* 19 (2013) 195–202.

[38] N. Ahn, Effects of diacrylate monomers on the mechanical properties of polymer concrete made with wet aggregates, *J. Appl. Polym. Sci.* 94 (2004) 1077–1085. https://doi.org/10.1002/app.13407.

[39] N. Ahn, Moisture sensitivity of polyester and acrylic polymer concretes with metallic monomer powders, *J. Appl. Polym. Sci.* 107 (2008) 319–323. https://doi.org/10.1002/app.

4 Polymer Concrete Standards

Abstract

Acquisition and improvement of various properties of polymer concrete (PC) as a product with wide applications has always been considered. To estimate and analyze the various properties of PC, experimental methods are used, which are generally destructive tests. Like polymer composites and cement concrete, the properties of PC can be evaluated according to existing published standards. The present chapter presents the existing standards for examining the properties of PCs. In addition to specific standards for PC, many studies have used standards for cement concrete and polymer composite to determine the properties of PC. These standards are collected and classified based on physical, mechanical, thermal, and chemical properties. Also, some non-destructive testing methods that have been used to determine the properties of PCs will be mentioned.

4.1 DESTRUCTIVE TESTS

Polymer concrete (PCs) are almost implemented in the structures and construction industry. Therefore, some characteristics such as mechanical strength, thermal properties, and chemical resistance are more important than others. Some properties like chemical resistance and physical characteristics of PCs are affected by the character

TABLE 4.1
International standards for the test methods of polymer concrete

Standards	Subjects
ASTM C267	Chemical resistance
ASTM C413	Absorption
ASTM C531	Linear shrinkage and coefficient of thermal expansion
ASTM C579	Compressive strength
ASTM C580	Flexural strength and modulus of elasticity
ASTM C905	Apparent density
DIN 51290-1	Terminology
DIN 51290-2	Testing of binders, fillers, and reactive resin compounds
DIN 51290-3	Testing of separately manufactured specimens
DIN 51290-4	In-process testing and testing of final parts
JIS A 1181	Test methods for PC
ACI 548.7-04	Load capacity
RILEM 113-CPT	Recommendations for test methods

DOI: 10.1201/9781003326311-4

TABLE 4.2
Some standards used to determine the properties of polymer concrete

Properties category	Evaluated properties	Standards
Physical properties	Bulk density	ASTM C138
	Apparent density	ASTM C905
	Absorption	ASTM C413
	Water absorption	ASTM D570
	Water absorption	SR EN ISO 10545-3:1999
	Water resistance	EN 12390-8
	Linear shrinkage	ASTM C531
	Drying shrinkage	ASTM C596
	Flowability	ASTM C1437
	Flowability	ASTM C230-90
	Flowability	ASTM C109-90
	Flowability	JIS R 5201
Mechanical properties	Compressive strength	ASTM C579
	Compressive strength	ASTM C109
	Compressive properties	ASTM D695
	Compressive strength	ASTM C39/C39M
	Compressive strength	RILEM standard CPT PC-2
	Compressive strength	Australian Standards AS1012.8.1
	Compressive strength	EN 12390-3
	Flexural strength	ASTM C580
	Flexural strength	ASTM C293
	Flexural strength	ASTM D790
	Flexural strength	ASTM C78-02
	Flexural strength	AS 1012.11
	Flexural strength	RILEM TC 113-CPT (PC-7)
	Flexural strength	RILEM CPT PCM-8 standard
	Flexural strength	ISO 178
	Flexural strength	DIN 1048-5
	Splitting tensile strength	ASTM C496
	Tensile strength	ASTM D638
	Tensile strength	ASTM D3967
	Tensile strength	ISRM suggestion
	Tensile strength	DIN 1048-5
	Fracture toughness	RILEM TC50-FMC
	Fracture toughness	ACI 446 (ACI 2009a)
	Fracture toughness	ASTM D5045-99
	Impact strength (Izod)	ASTM D256
	Durometer hardness	ASTM D2240
	Bond strength (slant shear)	ASTM C882
	Pull-off bonding strength	ASTM D4541
	Pull-off bonding strength	ASTM D7234-12
	Pull-off bonding strength	ASTM C1583/C1583M
	Fatigue strength	AASHTO T321-07

Properties category	Evaluated properties	Standards
	Abrasion resistance	ASTM C1138
	Recommendations for test methods	RILEM 113-CPT
	Guide for the use of polymers in concrete	ACI 548.1-9
	Test methods for PC	JIS A 1181
	Mechanical testing	DIN 51290-1, 2, 3, 4
Thermal properties	Coefficient of thermal expansion (CTE)	ASTM C531
	Coefficient of thermal expansion (CTE)	ASTM E228-85
	Coefficient of thermal expansion (CTE)	SR EN ISO 10545-8:2000
	Coefficient of thermal expansion (CTE)	RILEM TC/113 PC-13
	Thermal conductivity	STAS 5912-89
	Fire resistance	ASTM E1354
	Flammability	ASTM D635-97
	Rapid freezing and thawing resistance	ASTM C666
	Freeze–thaw	RILEM TC117-FDC
Chemical properties	Chemical resistance	ASTM C267
	Chemical reagents resistance	ASTM D543

of the implemented resin more thoroughly than by the aggregate type and content. Like cement concrete or polymers and polymer composites, standard methods are used to determine the properties of PC by experimental methods. Most experimental investigation on PC has been done using destructive testing (DT) methods. However, few studies have attempted to determine properties by non-destructive testing (NDT) methods. There are some international standards for the test methods of PCs that are listed in Table 4.1. The history of some of these standards was briefly described in Section 1.2.

In addition to the mentioned standards that are frequently used in research, some other standards related to testing polymers, polymer composites, or cement concretes have also been used in previous research. Table 4.2 shows the standard test methods to evaluate the properties of PC.

As shown in Table 4.2, several standards have been used by researchers to obtain the property of PC. Some parameters, such as specimen dimensions, differ in each standard. For example, standard ASTM C579 offers a cubic specimen for compressive strength testing, while standard ASTM C39/C39M offers a cylindrical one. However, both of them follow the fundamental theories of mechanical engineering [1].

4.2 NON-DESTRUCTIVE TESTS

In addition to the test methods mentioned in this book, which were often destructive tests, some properties of PCs can also be evaluated using NDT. However, only a few studies investigated the properties of PCs using NDT methods. For example, hardness, skid resistance, acoustic impact-echo, and VEI NDTs were conducted on the PC implemented in Bridge deck [2].

Ultrasonic pulse velocity (UPV) is one of the NDT methods that can provide various properties of PC. Factors such as the amount and type of resin, aggregates, and fillers, as well as the grain size distribution of aggregates and fillers affect the propagation of ultrasonic waves in PC [3]. UPV is not dependent to the shape and size of the sample, regardless of the frequency used [4]. This method can be used to study some properties like compressive strength [5], dynamic modulus and homogeneity of PC pre-fabricated parts and also to estimate its adhesion to cement concrete layers [6]. Another NDT procedure, ultrasonic echo method, can be implemented for determining the PC overlay thickness [6]. Impact resonance test was implemented for determining the dynamic modulus and dynamic shear modulus [7].

REFERENCES

[1] L.E. Hing, *Application of Polymer in Concrete Construction*, University of Technology Malaysia, 2008.
[2] R.J. Stevens, W.S. Guthrie, J.S. Baxter, B.A. Mazzeo, Field evaluation of polyester-polymer concrete overlays on bridge decks using nondestructive testing, *J. Mater. Civ. Eng.* 33 (2021) 4021155. https://doi.org/10.1061/(ASCE)MT.1943-5533.0003810.
[3] K. Zalegowski, Assessment of polymer concrete sample geometry effect on ultrasonic wave velocity and spectral characteristics, *Materials (Basel).* 14 (2021). https://doi.org/10.3390/ma14237200.
[4] S.K. Mantrala, C. Vipulanandan, Nondestructive evaluation of polyester polymer concrete, *Mater. J.* 92 (1995) 660–668.
[5] M.M. Abed, M.H. Almaamori, Z.J.A. Amer, Compression behavior of polymer concrete by using destructive and ultrasonic wave test at 26 khz, *Acad. Res. Int.* 4 (2013) 168–176.
[6] R. Kažys, O. Tumšys, D. Pagodinas, Ultrasonic detection of defects in strongly attenuating structures using the Hilbert–Huang transform, *NDT e Int.* 41 (2008) 457–466. https://doi.org/10.1016/j.ndteint.2008.03.006.
[7] J.D. Reis, R. De Oliveira, A.J.M. Ferreira, A.T. Marques, A NDT assessment of fracture mechanics properties of fiber reinforced polymer concrete, *Polym. Test.* 22 (2003) 395–401. https://doi.org/10.1016/S0142-9418(02)00120-4.

5 Physical Properties of Polymer Concrete

ABSTRACT

Although the physical properties of polymer concrete (PC) have not been considered in many studies, it is necessary to investigate them due to the direct or indirect effect of these properties on the strength, quality, and durability of PC. Experimental assessment demonstrates that the physical properties of PC depend on the ingredients, the composition of the materials, and the method of construction. Some physical characteristics of PC like density, porosity, ductility, water absorption, water resistance, flowability, and shrinkage, which were evaluated in the literature, are discussed in this chapter. These properties directly or indirectly affect the strength, quality, and durability of PC. Recognition and optimization of parameters affecting the mentioned properties can be useful in improving the physical properties of PC.

5.1 DENSITY

Like ordinary cement concretes, the density of polymer concretes (PCs) is especially important in structural applications. The density of PCs is usually lower than 2,000 kg/m^3 [1], and it is considered lightweight concrete. ASTM C905-01 [2] is a specific standard for determining the density of PC. Bulk density, which considers both the solids and the pore space, is determined according to ASTM C138 [3]. The bulk density is evaluated by dividing the total mass of concrete by the volume of the concrete [4]. The bulk density with 70 wt% river sand and 30 wt% polyester is about 1,600 kg/m^3 [4].

The use of insufficient amounts of resin in PC reduces the density due to the formation of the porous structure and large cavities. By implementing more binder wt%, resin fills the cavities, and therefore the density increases, as demonstrated in Figure 5.1 [5]. Afterward, increasing polymer weight content causes to decrease in the bulk density value due to the lesser density of polymer binder [4].

As illustrated in Figure 5.2, the use of equal amounts of fine and coarse aggregates in the composition of PC gives the highest density. This result can be attributed to the filling of the empty space between the coarse aggregate with fine aggregates and as a result a more compacted concrete composition [6].

As demonstrated in Figure 5.3, curing age only has a slight effect on the density of the various epoxy PC. The most important increase is related to the PC with 13 wt% resin, which does not exceed 0.89%. Such increment was imputed to improve the cross-linking density of the polymer [5].

Efforts have been conducted to introduce lightweight PC by implementing lightweight aggregates [7] and lightweight foam resin [8].

DOI: 10.1201/9781003326311-5

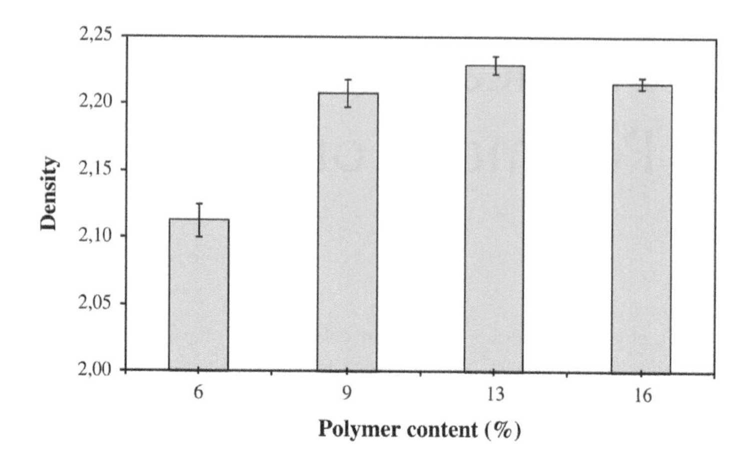

FIGURE 5.1 Variation of the density of epoxy PC in terms of polymer content [5]

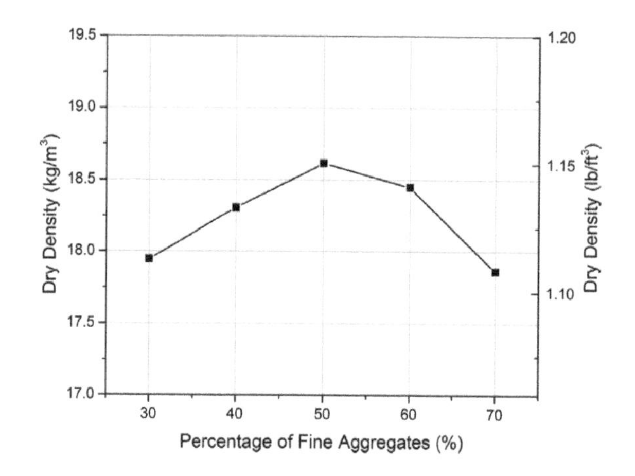

FIGURE 5.2 Dry density of epoxy PC with the variation of fine aggregate [6]

5.2 POROSITY

From an engineering point of view, the porosity of the concrete should be minimized. But its complete removal is out of reach. Porosity directly affects the strength, mass, and durability of PC. Porosity in PC can be greatly minimized by proper molding of specimens.

Generally, PCs have less porosity than cement concretes due to the lack of water in their structure. The low porosity of PCs causes improvement in chemical stability (resistance to carbonation, chloride ion, and oxygen penetration). Thus, they can be implemented as an anticorrosive material. Intrusion porosimetry is a commonly extended test method to determine pore size distribution in concrete faster and precisely than any other procedure [9]. In this method, intrusion at low (0.34 MPa) and

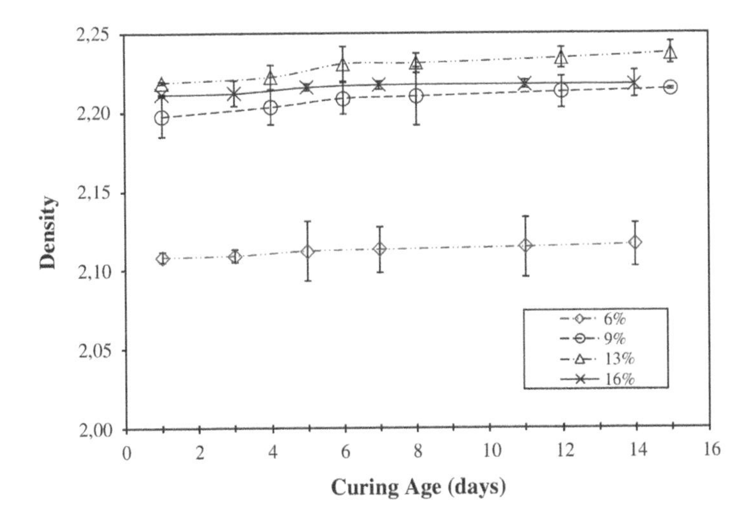

FIGURE 5.3 The variation of density in different curing ages for the various polymer ratios [5]

high (414 MPa) pressures of the mercury in the porosity of the specimen is obtained, and pore volume of the specimen is derived by calculating the volume of mercury staying in the penetrometer stem [10].

It was reported that increasing the polymer amount from 6 to 13 wt% decreased the porosity of the PC. It was mentioned that populations of capillary pores are founded only for PCs, including less than 9 wt% polymer binder. This phenomenon is mainly because of the absence of sufficient resin to bond the aggregates, and therefore, the risk of voids enhances [5]. A similar result was reported in Ref. [11]. As can be seen in Figure 5.4, the results of both mercury intrusion porosimeter and the calcination are quite similar. By increasing the polymer ratio, the porosity decreased.

Figure 5.5 shows the porosity of epoxy PC for various micro filler contents. Changing the ratio of resin to micro filler from 100:0 to 40:60 (increasing the micro filler content) increased the porosity by about 328% [12].

5.3 DUCTILITY

Generally, ductility is used to determine the ability of a material to deform plastically and absorb energy during deformation. In the case of concrete materials, ductility is of great importance in terms of strength and serviceability. At their near-maximum load-carrying capacity, concrete structures should be capable of undergoing large deflections to prevent total collapse. It was reported that the ductility of pc is greater than that of ordinary Portland cement (OPC) concrete [13]. There is no agreement on the finest procedure of determining ductility. The most widely admitted description of ductility is the ratio of curvature of the section at the end and the beginning of the plastic plateau of the load–curvature curve [13,14]. The ductility factor μ is defined by Eq. (5.1) where ε_1 is the approximated limit of the elastic strain point, which is the

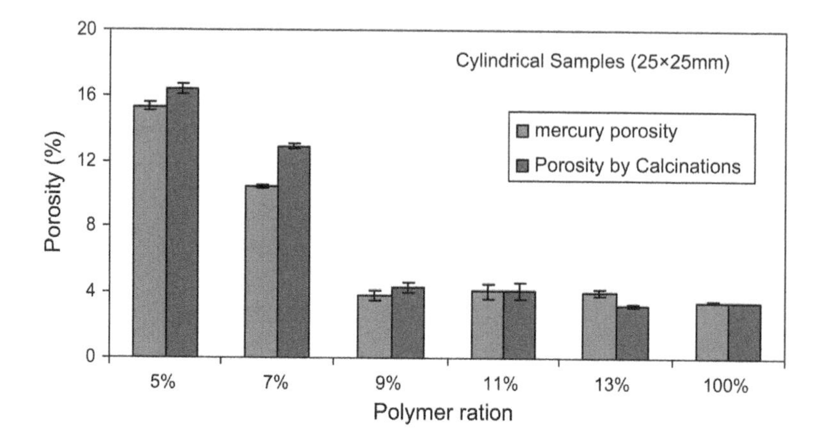

FIGURE 5.4 The variation of the porosity of epoxy PC with the polymer ratios [11]

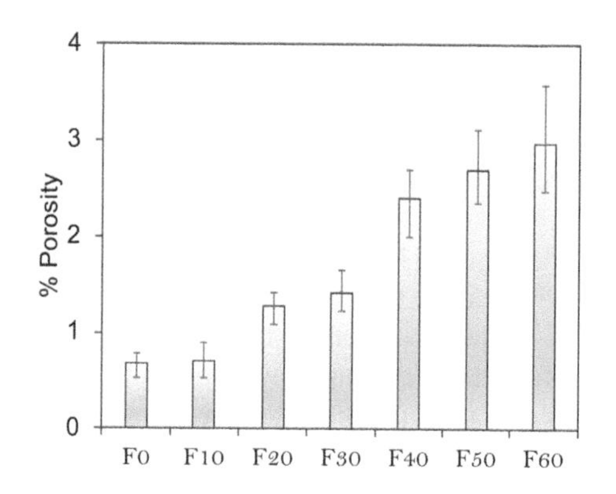

FIGURE 5.5 The variation of the porosity of epoxy PC with the micro filler content [12]

crossing point of the peak stress and the extension line of the linear elastic region, and ε_2 is the strain at 85% of the peak stress in the plastic region [15,16]. As illustrated in Figure 5.6, increasing the environmental temperature increases the ductility factor [16].

$$\mu = \frac{\varepsilon_2}{\varepsilon_1} \tag{5.1}$$

Incorporating fly ash micro filler into vinyl ester and epoxy PC reduced the ductility level. So proper consideration should be given for PCs, including fly ash used in structural applications [15]. Applying alumina nanoparticles (ANPs) lead to producing epoxy PC with superior ductility. ANPs increased tensile strain at failure (as a numeric measure of ductility) from 2.6 for neat PC to 4.9%. Increasing the ductility was attributed to the capability of ANPs to decrease polymer cross-linking intensity

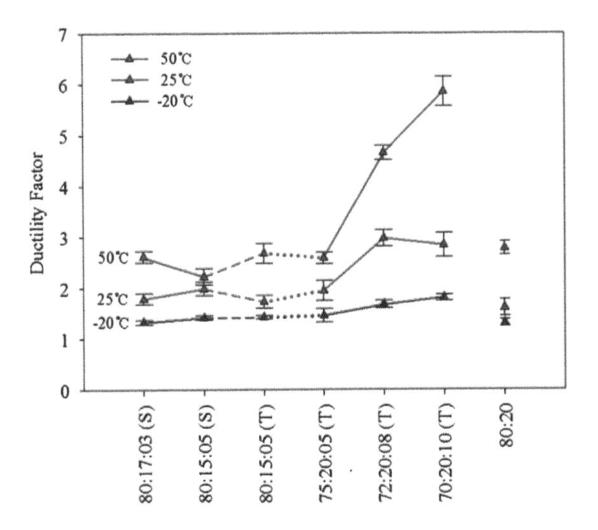

FIGURE 5.6 Ductility factor (m) of various PCs according to mix proportion and environmental temperature [16]

[17]. Increasing the resin-to-filler ratio is another approach for improvements in ductility. However, such an increase causes decrease in the flexural and compressive strengths of the PC up to 70% and 26%, respectively [18]. Also, implementing 0.001–0.018 wt% carboxyl (COOH) functionalized multi-walled carbon nanotubes (MWCNTs) maximizes epoxy PC ductility attaining an outstanding 5.5% tensile strain at failure, which is five times higher than neat PC [19].

5.4 WATER ABSORPTION

The water absorption rate of PCs depends on polymer binders, fillers, and aggregates. Based on the mixture, obtained water absorption of OPC concretes is two to four times higher than the PCs. Such low water absorption of PC is attributed to the negligible water absorbency of the implemented polymeric matrix. For determining the water absorption of PC according to ASTM D570 standard [20], the sample dried at 80±2°C for a specific time and was weighted. Then immersed in water at room temperature for 24 hours and weighed again. The water absorption rate is calculated as follows:

$$\text{Absorption rate } (\%) = \frac{W_1 - W_0}{W_0} \times 100 \tag{5.2}$$

where W_0 and W_1 are the sample weights before and after the water absorption test, respectively [21].

Increasing the polymer binder weight content causes a reduction in water absorption. More polymer binder surrounds micro filler and aggregates and prevents water from reaching the material. The addition of micro filler content with a low

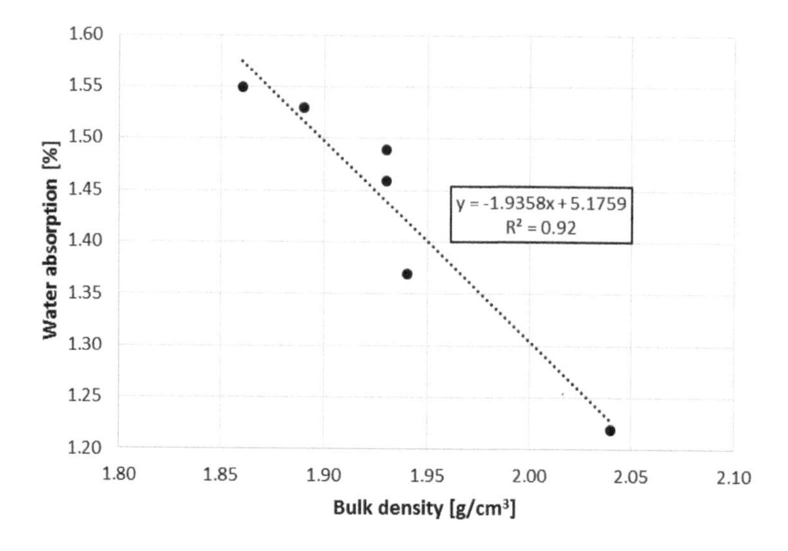

FIGURE 5.7 Water absorption of the PC with RGA as a function of bulk density [22]

absorbency rate decreases the water absorption rate of the PC. Micro filler bond with binder and aggregates, decrease the microvoids and avoid water penetration into the concrete [9].

Substituting the recycled glass aggregate (RGA) with the mixture of sand and gravel in polyester PC decreased the water absorption by 27%. The behavior of the water absorption of the PCs in terms of bulk density is illustrated in Figure 5.7. Increasing the bulk density declined the water absorbency of the PCs, which was ascribed to the low water absorption of RGA [22].

As resin content increases, water absorption of PC is decreased. However, when unsaturated orthophthalic polyester resin exceeds 12 wt%, water absorption exhibits a slight increase [23].

The epoxy PC exhibited a low water absorption of less than 1%, which prevented the intrusion of water and salts and avoided the corrosion of substrates. Generally, PCs have a low porosity that reverberates the increase in carbonation resistance, decreases oxygen and chemical ion permeation, and therefore improves anti-corrosion property. Thus, the PC can be implemented as a corrosion-resistance coating [24].

Mani et al. determined the influence of variables on water absorption of furan PC. It was reported that the lowest water absorption belonged to the PC inclusion of 7.5 wt% resin. By increasing the resin content, the micro filler content decreased, and thus, the number of microvoids that were not being filled increased. Therefore, the water absorption increased. Also, it can be attributed to the existence of microvoids produced during the curing of furan releasing water molecules [25].

The one-year immersion of the polyester PC in water (pH = 8) reduced the compressive strength from 66 MPa to 47.5 MPa [26]. On the other side, neat polyester specimens indicated a strength reduction of less than 5% after three years immersion in water. Therefore, water diffusion in the PC affected the binder–aggregate interface

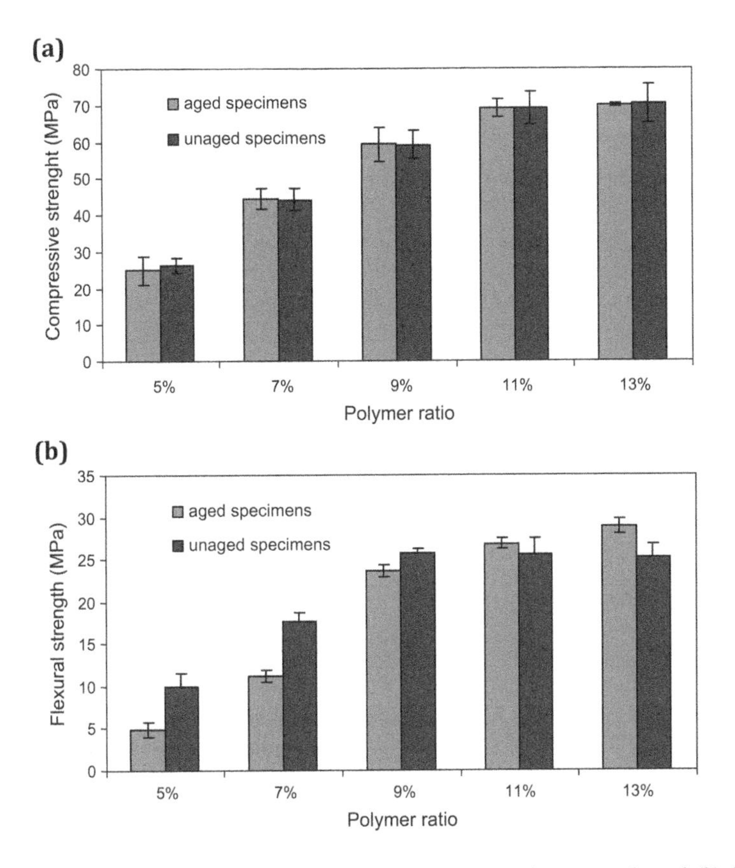

FIGURE 5.8 Effect of water absorption on the (a) compressive strength and (b) flexural strength of epoxy PC [11]

rather than the polymer. By increasing the total water amount uptake into the PC, the compressive and splitting tensile strengths decreased. However, the compressive strength was almost entirely recovered upon redrying the PC specimens. The authors introduced a simple model based on diffusion theory to determine the strength degradation rate for PC in water. The diffusion coefficient for PC in water was determined to be 2.5 × 10–11 m^2/s.

Figure 5.8 shows the effect of the water absorption on the compressive and flexural strengths of epoxy PC in terms of polymer ratio. The compressive strength values of the aged specimens (immersed in water) don't exhibit a significant reduction compared to the unaged specimens. However, the flexural strength of the specimens with low polymer content up to 9%, has shown a significant reduction after immersion in water. It can be attributed to the higher permeability and porosity of the specimens with the lower polymer content [11].

The epoxy PC represented more negligible water absorption than the polyester PC. But, both of them exhibited considerably lower water absorption than the OPC concrete [27]. Due to low water absorption, which prevents the intrusion of water and salts, PC is implemented as an anti-corrosion coating for Portland Cement Concrete and steel substrates [28]. Water absorption of the epoxy and polyester PC is greater in boiling water than in ambient temperature water. The implementation of the silane coupling agent reduced the water absorption of the polyester PC slightly. The negligible water absorption value around the water freezing temperature corroborated considerable freeze–thaw resistance for PC compared to the OPC concrete [27]. Methyl methacrylate (MMA)-based PC demonstrated a little water absorption of 0.039–0.089% after the first day and 0.015–0.330% after eight days [29].

Literature [30,31] evaluated the depth of water penetration based on EN 12390-8 method at the age of 28 and 90 days. Both the unsaturated polyester resin (NR) and the modified polyester resin (MR) PCs have negligible water penetration at different ages, compared to normal concrete (NC) [30].

For determining the hot water absorption of PC according to SR EN ISO 10545-3:1999, the specimens were immersed in boiling water at 100°C for 2 hours and then cooled in water for a further 2 hours and 15 minutes. The mean value of hot water absorption was obtained 0.627% [1].

5.5 WATER RESISTANCE

Water resistance is the maximum water pressure that doesn't cause leakage and is usually determined for OPC concrete. The corrosive effect of aqueous medium on PC appears as an alteration in its specifications and structure without compromising the integrity or by destroying the material. Water penetrates through voids and tiny capillaries between macromolecules and irreversibly alters the chemical structure of the PC [32]. As a result of these changes, many properties, including the strength of PC, will undergo massive changes. Therefore, the performance of PC largely depends on the humidity of the environment.

A water resistance test was performed according to EN 12390-8 [33]. Water resistance of the polyester PC was investigated by applying hydrostatic pressure of 0.2 MPa for 24 hours and increasing the water pressure by 0.2 MPa every 24 hours on the one side of a cubic specimen until leakage signs became apparent [22]. The water resistance of all PC samples was obtained higher than W6. The hydrostatic pressure of 6 MPa did not cause any leakage. The good water resistance of the PC was attributed to the highly waterproof polymer binder. Reference [32] reviewed several research papers on the effects of moisture on the physical and mechanical properties of PCs. Although proximity to moisture or aqueous media reduces some of the properties of PCs, these changes are much less than cement concretes.

Figure 5.9 shows compressive stress–strain behavior of the epoxy PC after long-term exposure in water. It is observed that being in water has increased the compressive strength of PC. This increase is 26% in the first four months, and the changes after that are not significant [12].

5.6 FLOWABILITY (WORKABILITY)

A flowability test is conducted to investigate the workability of the PC mixture. Since there is no standard test method for PC flowability, authors carried out the

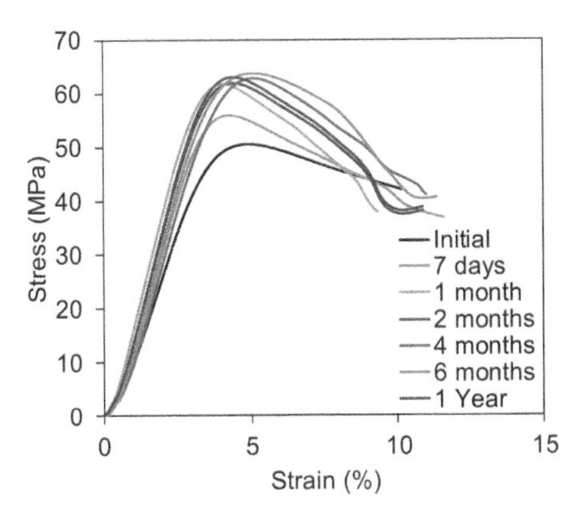

FIGURE 5.9 Compressive stress–strain diagram of the epoxy PC after up to 365 days exposure in water [12]

test according to Slump test [34], ASTM C 1437 [17,35–37] ASTM C 230-90 [38], ASTM C 109-90 [38], and JIS R 5201 [39]. The viscosity of some resins increases with decreasing temperature and therefore, the workability decreases.

Commercially available unsaturated polyester resins are highly viscous and a diluent agent is frequently added to achieve the required flowability. Authors [37,40] implemented MMA to remove the viscosity problem of UP resin. Applying more amount of MMA increased the workability of PC. The addition of 40 wt% MMA significantly enhanced the flow value from 105 mm to 159.6 mm.

Ahn [38] studied the influences of three metallic monomer powders, including zinc diacrylate (ZDA), zinc dimethacrylate (ZMA), and calcium diacrylate (CDA), on the performance of fresh acrylic and polyester PC. All the above-mentioned powders did not dissolve in MMA monomer and did not influence the workability of MMA PC. On the other side, ZDA and ZMA increased the workability of polyester PC, but CDA reduced its workability.

The addition of micro and nanofillers can change the flowability of the PC mixture, which influences mechanical performance and also imposes difficulties and limitations with in-situ applications. Incorporating the MWCNTs [19,35,36], alumina nanoparticles [17] silica nanoparticles [41] reduced the flowability of the PC because of the increased viscosity of the mixture. However, ground palm oil fuel ash (GPOFA) had enhanced the flowability of the PC [39]. As presented in Figure 5.10, the introduction of 1.5 wt% MWCNTs decreased the flowability of epoxy PC by 66% [36].

5.7 SHRINKAGE

Low shrinkage property is important, especially in precast applications. Shrinkage strain can significantly influence the dimension of PC members and makes problems with demolding and use of them. Unlike cement-based concretes, which experience both short-term (plastic) and long-term (drying) shrinkages because of water vaporization, PCs only undergo short-term shrinkage during the resin polymerization [42].

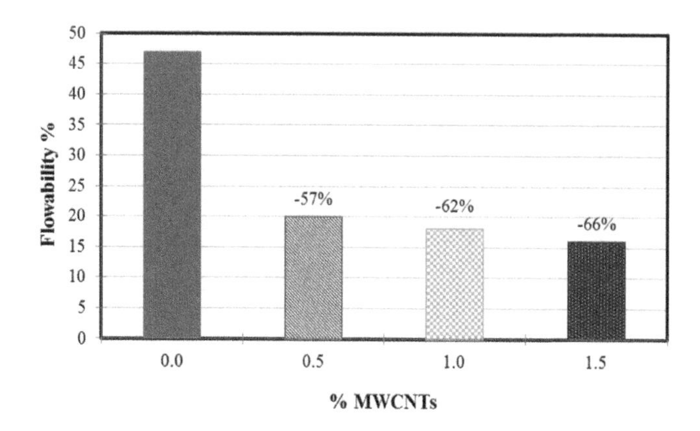

FIGURE 5.10 Effect of MWCNT wt% on the flowability of the epoxy PC [36]

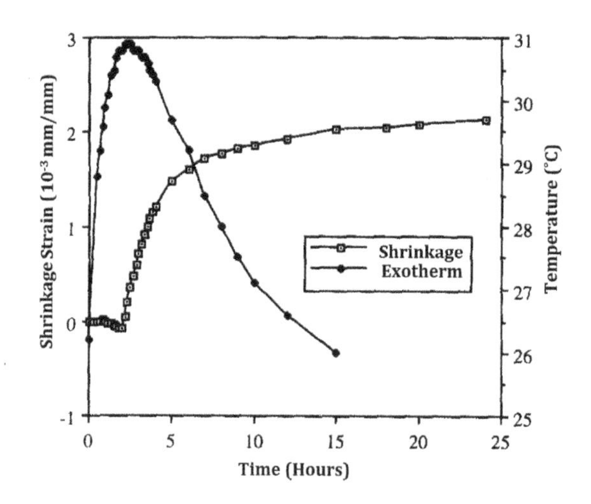

FIGURE 5.11 Typical shrinkage and exotherm [42]

Setting shrinkage of PC is influenced by the type, and weight content of the catalyst and accelerator added, as well as the curing temperature. The PC obtained 90% of its ultimate setting shrinkage at the age of 3 hours [43]. In particular applications such as crack repair of cementitious structures or in the surfacing of roads and pavements, where shrinkage of PC cannot be accommodated, the use of shrinkage controlling agents is essential [27]. For example, the setting shrinkage of the unsaturated polyester PC decreased from 73 × 10^{-4} to 23 ×10^{-4} by adding 40 wt% MMA [37]. Applying $CaCO_3$ micro filler decreased the setting shrinkage of the polyester PC [27]. It was shown that the shrinkage of the PC can be reduced to zero or can even be negative (i.e., expansion), depending on the applied shrinkage reducing agent [27]. As illustrated in the typical shrinkage-exotherm curve of the PC (see Figure 5.11), most of the shrinkage occurs after the incidence of the peak exotherm and within the first 8 hours after mixing and stops after 24 hours [42].

REFERENCES

[1] L. Agavriloaie, S. Oprea, M. Barbuta, F. Luca, Characterisation of polymer concrete with epoxy polyurethane acryl matrix, *Constr. Build. Mater.* 37 (2012) 190–196. https://doi.org/10.1016/j.conbuildmat.2012.07.037.

[2] ASTM C905-01, Standard test methods for apparent density of chemical-resistant mortars, grouts, monolithic surfacings, and polymer concretes, *ASTM Int. West Conshohocken, PA.* (2012) 1–4.

[3] ASTM C138/C138M-17a, Standard test method for density (unit weight), yield, and air content (gravimetric) of concrete, *ASTM Int. West Conshohocken, PA.* (2017) 1–6.

[4] A.M. Hameed, M.T. Hamza, Characteristics of polymer concrete produced from wasted construction materials, *Energy Procedia.* 157 (2019) 43–50. https://doi.org/10.1016/j.egypro.2018.11.162.

[5] O. Elalaoui, E. Ghorbel, V. Mignot, M. Ben Ouezdou, Mechanical and physical properties of epoxy polymer concrete after exposure to temperatures up to 250°C, *Constr. Build. Mater.* 27 (2012) 415–424. https://doi.org/10.1016/j.conbuildmat.2011.07.027.

[6] V. Toufigh, M. Hosseinali, S. Masoud, Experimental study and constitutive modeling of polymer concrete's behavior in compression, *Constr. Build. Mater.* 112 (2016) 183–190. https://doi.org/10.1016/j.conbuildmat.2016.02.100.

[7] H. Sanaei Ataabadi, A. Zare, H. Rahmani, A. Sedaghatdoost, E. Mirzaei, Lightweight dense polymer concrete exposed to chemical condition and various temperatures: An experimental investigation, *J. Build. Eng.* 34 (2021) 101878. https://doi.org/https://doi.org/10.1016/j.jobe.2020.101878.

[8] M.H. Niaki, M.G. Ahangari, A. Fereidoon, Mechanical properties of reinforced polymer concrete with three types of resin systems, *Proc. Inst. Civ. Eng. – Constr. Mater.* 0 (2022) 1–24. https://doi.org/10.1680/jcoma.21.00060.

[9] E. Hwang, J. Kim, S. Park, Physical properties of polyester polymer concrete composite using RCSS fine aggregate, *Adv. Mater. Res.* 687 (2013) 219–228. https://doi.org/10.4028/www.scientific.net/AMR.687.219.

[10] J. Toma, Reinforced polymer concrete: Physical properties of the matrix and static/dynamic bond behaviour, *Cem. Concr. Compos.* 27 (2005) 934–944. https://doi.org/10.1016/j.cemconcomp.2005.06.004.

[11] M. Haidar, E. Ghorbel, H. Toutanji, Optimization of the formulation of micro-polymer concretes, *Constr. Build. Mater.* 25 (2011) 1632–1644. https://doi.org/http://dx.doi.org/10.1016/j.conbuildmat.2010.10.010.

[12] W. Ferdous, A. Manalo, H.S. Wong, R. Abousnina, O.S. Alajarmeh, Y. Zhuge, P. Schubel, Optimal design for epoxy polymer concrete based on mechanical properties and durability aspects, *Constr. Build. Mater.* 232 (2020) 117229. https://doi.org/10.1016/j.conbuildmat.2019.117229.

[13] H. Abdel-Fattah, M.M. El-Hawary, Flexural behavior of polymer concrete, *Constr. Build. Mater.* 13 (1999) 253–262. https://doi.org/10.1016/S0950-0618(99)00030-6.

[14] B. Jo, S. Park, K. Lee, Prediction of stress – strain relationship for polyester polymer concrete using recycled polyethylene terephthalate under compression, *Adv. Cem. Res.* 7605 (2008) 151–159. https://doi.org/10.1680/adcr.2007.00032.

[15] W. Lokuge, T. Aravinthan, Effect of fly ash on the behaviour of polymer concrete with different types of resin, *Mater. Des.* 51 (2013) 175–181. https://doi.org/10.1016/j.matdes.2013.03.078.

[16] I. Roh, K. Jung, S. Chang, Y. Cho, Characterization of compliant polymer concretes for rapid repair of runways, *Constr. Build. Mater.* 78 (2015) 77–84. https://doi.org/10.1016/j.conbuildmat.2014.12.121.

[17] M. Emiroglu, A.E. Douba, R.A. Tarefder, U.F. Kandil, M.R. Taha, New polymer concrete with superior ductility and fracture toughness using alumina nanoparticles, *J. Mater. Civ. Eng.* 29 (2017) 1–9. https://doi.org/10.1061/(ASCE)MT.1943-5533.0001894.

[18] W. Ferdous, A. Manalo, T. Aravinthan, G. Van Erp, Properties of epoxy polymer concrete matrix: Effect of resin-to-filler ratio and determination of optimal mix for composite railway sleepers, *Constr. Build. Mater.* 124 (2016) 287–300. https://doi.org/https://doi.org/10.1016/j.conbuildmat.2016.07.111.
[19] A. Douba, M. Emiroglu, U.F. Kandil, M.M.R. Taha, Very ductile polymer concrete using carbon nanotubes, *Constr. Build. Mater.* 196 (2019) 468–477. https://doi.org/10.1016/j.conbuildmat.2018.11.021.
[20] ASTM D570-98, Standard test method for water absorption of plastics, *ASTM Int.* (2018) 1–4. https://doi.org/10.1520/D0570-98R18.
[21] J. Wang, Q. Dai, S. Guo, R. Si, Mechanical and durability performance evaluation of crumb rubber-modified epoxy polymer concrete overlays, *Constr. Build. Mater.* 203 (2019) 469–480. https://doi.org/10.1016/j.conbuildmat.2019.01.085.
[22] P. Ogrodnik, Physico-mechanical properties and microstructure of polymer concrete with recycled glass aggregate, *Materials (Basel).* 11 (2018) 1–15. https://doi.org/10.3390/ma11071213.
[23] F. Carrión, L. Montalbán, J.I. Real, T. Real, Mechanical and physical properties of polyester polymer concrete using recycled aggregates from concrete sleepers, *Sci. World J.* 2014 (2014) 1–10. https://doi.org/http://dx.doi.org/10.1155/2014/526346.
[24] K. Shi-cong, P. Chi-sun, A novel polymer concrete made with recycled glass aggregates, fly ash and metakaolin, *Constr. Build. Mater.* 41 (2013) 146–151. https://doi.org/10.1016/j.conbuildmat.2012.11.083.
[25] M. Muthukumar, D. Mohan, Studies on furan polymer concrete, *J. Polym. Res.* 12 (2005) 231–241. https://doi.org/10.1007/s10965-004-3206-7.
[26] S. Mebarkia, C. Vipulanandan, Mechanical properties and water diffusion in polyester polymer concrete, *J. Eng. Mech.* 121 (1995) 1359–1365. https://doi.org/https://doi.org/10.1061/(ASCE)0733-9399(1995)121:12(1359).
[27] P. Mani, A.K. Gupta, S. Krishnamoorthy, Comparative study of epoxy and polyester resin-based polymer concretes, *Int. J. Adhes. Adhes.* 7 (1987) 157–163. https://doi.org/10.1016/0143-7496(87)90071-6.
[28] M.E. Tawfik, S.B. Eskander, Polymer concrete from marble wastes and recycled poly(ethylene terephthalate), *J. Elastomers Plast.* 38 (2006) 65–79. https://doi.org/10.1177/0095244306055569.
[29] L. Trykoz, S. Kamchatnaya, O. Pustovoitova, A. Atynian, O. Saiapin, Effective waterproofing of railway culvert pipes, *Balt. J. Road Bridg. Eng.* 14 (2019) 473–483. https://doi.org/https://doi.org/10.7250/bjrbe.2019-14.453.
[30] M. Jamshidi, A. Reza, Modified polyester resins as an effective binder for polymer concretes, *Mater. Struct.* 45 (2012) 521–527. https://doi.org/10.1617/s11527-011-9779-9.
[31] M. Jamshidi, A.R. Pourkhorshidi, A comparative study on physical/mechanical properties of polymer concrete and portland cement concrete, *Asian J. Civ. Eng.* 11 (2010) 421–432.
[32] Oleg Figovsky, D. Beilin, *Advanced Polymer Concretes and Compounds*, 1st ed., CRC Press, Boca Raton, FL, 2013. https://doi.org/https://doi.org/10.1201/b16237.
[33] BS EN 12390-8:2009, Testing hardened concrete, Part 8: Depth of penetration of water under pressure, *Eur. Comm. Standarization*, Brussels, Belgium. (2009).
[34] N. Ahn, D.K. Park, J. Lee, M.K. Lee, Structural test of precast polymer concrete, *J. Appl. Polym. Sci.* 114 (2009) 1370–1376. https://doi.org/10.1002/app.
[35] S.M. Daghash, E.M. Soliman, U.F. Kandil, M.M. Reda Taha, Improving impact resistance of polymer concrete using CNTs, *Int. J. Concr. Struct. Mater.* 10 (2016) 539–553. https://doi.org/10.1007/s40069-016-0165-4.
[36] S.M. Daghash, *New Generation Polymer Concrete Incorporating Carbon Nanotubes*, The University of New Mexico, Thesis for Master of Science, 2013.

[37] K.-S. Yeon, N.J. Jin, J.H. Yeon, Effect of methyl methacrylate monomer on properties of unsaturated polyester resin-based polymer concrete, in: M.M.R. Taha (Ed.), *Int. Congr. Polym. Concr. (ICPIC)*, Springer, Cham, Washington, DC, 2018: pp. 165–171. https://doi.org/https://doi.org/10.1007/978-3-319-78175-4_19.

[38] N. Ahn, Influences of metallic polymeric materials on the properties of fresh polyester and acrylic polymer concrete, *J. Appl. Polym. Sci.* 99 (2006) 2337–2343. https://doi.org/10.1002/app.22846.

[39] J. Mirza, N. Hafizah, A. Khalid, M. Warid, Effectiveness of palm oil fuel ash as micro-filler in polymer concrete, *J. Teknol. Full.* 16 (2015) 75–80.

[40] R. Drochytka, J. Hodul, Experimental verification of use of secondary raw materials as fillers in epoxy polymer concrete, in: M.M.R. Taha (Ed.), *Int. Congr. Polym. Concr. (ICPIC)*, Springer, Cham, Washington, DC, 2018: pp. 135–141. https://doi.org/https://doi.org/10.1007/978-3-319-78175-4_15.

[41] A. Douba, M. Genedy, E. Matteo, U.F. Kandil, M.M.R. Taha, The significance of nanoparticles on bond strength of polymer concrete to steel, *Int. J. Adhes. Adhes.* 74 (2017) 77–85. https://doi.org/10.1016/j.ijadhadh.2017.01.001.

[42] K.S. Rebeiz, Time-temperature properties of polymer concrete using recycled PET, *Cem. Concr. Compos.* 17 (1995) 119–124. https://doi.org/10.1016/0958-9465(94)00004-I.

[43] Y. Kyu-Seok, Y. Jung-Heum, Setting shrinkage, coefficient of thermal expansion, and elastic modulus of UP-MMA based polymer concrete, *J. Korea Concr. Inst.* 24 (2012) 491–498. https://doi.org/http://dx.doi.org/10.4334/JKCI.2012.24.4.491.

6 Mechanical Properties of Polymer Concrete

Abstract

Concrete structures and prefabricated members are exposed to static and dynamic loads when used. Therefore, recognition of their mechanical behavior and properties is essential. In previous works, some of the mechanical properties of polymer concretes (PCs), like compressive strength, flexural strength, and tensile strength, are of greater interest than the others. But according to the type of application of these concretes, the determination of other mechanical properties has also been discussed. The present chapter evaluates some of the important mechanical properties of PCs as discussed in the literature. Compressive, flexural, tensile, and impact strengths, fracture toughness, bond and interfacial shear strength, fatigue strength, creep behavior, vibration and damping properties, and abrasion and wear resistance of PC are investigated and various parameters affecting these properties are discussed.

6.1 COMPRESSIVE STRENGTH

One of the most important properties of polymer concrete (PC) as a construction material is its compressive strength. In many researches, the effect of additives on compressive strength has been investigated as a main criterion of concrete strength. Different standard test methods were implemented in research works to determine the compressive strength of PC like ASTM C579 [1–3], ASTM C39/C39M [4,5], ASTM: D695 [6,7], RILEM standard CPT PC-2 [8], Australian Standards AS1012.8.1 [9], EN 12390 [10,11] and so on. From a comparative study, PCs have a much higher compressive strength up to five times than ordinary Portland cement (OPC) concretes and up to three times than durable concretes, after seven days of curing [12,13].

According to a research study conducted in literature [14], the compressive strength of the PC has been reached to be close to 130 MPa. As studied in Chapter 2, different parameters such as resin, fillers, and aggregates amount and size play a crucial role in the compressive property of PC.

Authors derived a prediction model for compressive strength of vinyl ester resin PC at a young age using the maturity method [15]. The appropriate model was DR-HILL, which is expressed as follows:

$$S = \alpha + \frac{\theta M^{\eta}}{\kappa^{\eta} + M^{\eta}} \tag{6.1}$$

where S is the compressive strength (MPa), M is the maturity index (°C·hours or °C·days), and α, θ, η, and κ are parameters.

DOI: 10.1201/9781003326311-6

Yeon [16] studied the stress–strain relation and the length effect of the flexural compression member of UP-MMA PC. A series of C-shaped PC samples were subjected to flexural compressive load. Test results indicate that PCs have a strong length-dependent size effect.

Based on the Brazilian standard NBR 5739 [17], there are five possible failure patterns for cylindrical cement-based concrete samples, including (1) conic, (2) shear-conic, (3) shear diagonal, (4) shear, and (5) column-like. In the case of PCs, as reported in the literature, most of the failure patterns occur according to the shear pattern. Figure 6.1 represents the failure modes of basalt fiber-reinforced epoxy PC specimens under compressive load [18].

6.2 FLEXURAL STRENGTH

Flexural strength is one of the main properties of PCs which were discussed in numerous papers. Various standard test methods including both three-point and four-point bending tests were utilized in literatures such as ASTM C580 [1,19], ASTM C293 [20], ASTM C78-02 [21], Australian Standards AS 1012.11 [9], RILEM TC 113-CPT (PC-7) [22,23], RILEM CPT PCM-8 standard [24,25], ISO 178 [26], and DIN 1048-5 [12].

The flexural strength of the PC can reach up to 40 MPa. But in the case of fiber and nano-reinforced PC, this value exceeds 50 MPa [18]. PC has flexural strength up to 5.5 times more than OPC concrete and up to 3.5 times more than durable concrete [12].

Considering the mathematical relationships governing the compressive and the flexural strengths of epoxy/basalt PCs exhibited that they both follow a power law. The relevant empirical expression obtained was

$$f_{sp} = 1.5\, f_c^{0.42} \tag{6.2}$$

where f_{sp} and f_c are the experimental splitting tensile and compressive strengths, respectively, given in MPa.

FIGURE 6.1 Failure patterns of basalt fiber-reinforced epoxy PC specimens under compression load (a) shear, (b) column-like, and (c) shear-conic [18]

6.3 TENSILE STRENGTH

Low tensile strength is one of the major drawbacks of cementitious concrete. PCs implement polymer binders, and due to the acceptable tensile property of the polymers, the tensile strength of PCs is about four times more than the conventional cement concretes and about 2.5 times more than durable concrete [12,13].

Authors implemented different standard test methods such as ASTM C496/C496M [19,27], ASTM D638-14 [28], ASTM D3967 [26], ISRM suggestion [20], and DIN 1048-5 [12] for determining the tensile strength of PCs. Usually, because of the difficulty of direct tensile testing for concrete, indirect test procedures such as Brazilian disc sample exposed to diametral compression loading [29,30], rectangular beam exposed to flexural three or four-point bend loading [19,31], the split tensile test subjected to diametral compression loading [14,32], the semi-circular bend (SCB) specimen exposed to flexural three-point bend loading [33] are implemented. Among the methods mentioned above, the splitting tensile test is the most implemented procedure to determine the tensile strength of the PCs.

The splitting tensile test is an indirect tension test method, in which compressive load was exerted diametrically along the length of the concrete cylinder, and due to the indirect tensile stress generated by Poisson's effect, the specimen splits into two halves along the vertical plane as depicted in Figure 6.2. The splitting tensile strength generally is higher than the direct tensile strength but lower than the flexural strength [34,35] and calculated as follows:

FIGURE 6.2 Failure of PC specimen under the splitting tensile strength [26]

$$\sigma_t = \frac{2p_t}{\pi LD} \tag{6.3}$$

where p_t is the maximum exerted load, and L and D are the height and diameter of the specimen, respectively.

Studying the governing equation on the relationship of compressive and the splitting tensile strengths of an epoxy/basalt PC revealed that they both follow a power law according to the empirical expression as follows:

$$f_{sp} = 1.5\ f_c^{0.42} \tag{6.4}$$

where f_{sp} and f_c are the experimental values of splitting tensile strength and compressive strength, respectively, given in MPa [36]. The flexural strengths of the polyester, vinyl ester, and epoxy PCs have the same general behavior as the splitting tensile strength [26,37].

6.4 IMPACT STRENGTH

One of the main weaknesses of cementitious concretes is low impact resistance. Polymeric concrete has a higher impact resistance due to the polymer matrix. Minimizing the implemented resin in PC to decrease the cost of construction will reduce the impact strength. Therefore, it is necessary to examine the impact strength and provide solutions to improve that. Only a few research works concentrate on improving the impact strength of PC using additives.

In the literature [38], the Charpy impact test procedure was performed to calculate the impact properties of the PC. Figure 6.3 demonstrates the failure mode of a PC specimen in the Charpy test, which is combined flexure and shear fracture. Charpy impact test did not present any statistical difference with the recorded test energy. It is due to the great impact energy applied in the Charpy test. Thus, the Charpy impact test is not an appropriate procedure to determine the energy absorption of PC and is typically suitable for metals. Thus, the low-velocity impact test was performed, and

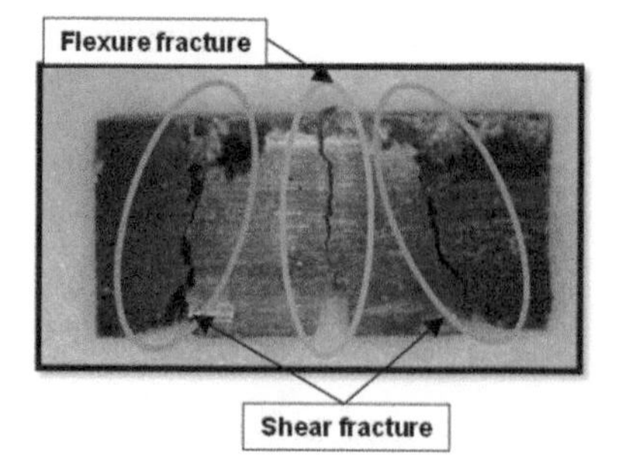

FIGURE 6.3 The failure mode of the PC specimen under the Charpy impact test [38]

9% increase in impact strength was recorded corresponding to 1 wt% multi-walled carbon nanotubes (MWCNTs).

Some researchers evaluated the impact strength of PCs using the Izod impact test method with the ASTM D256 standard [18,39]. The impact strength of an unreinforced epoxy PC was 3.13 KJ/m². Chopped basalt fiber increased impact strength significantly. The inclusion of 3 wt% basalt fiber enhanced the impact strength by more than four times higher than the neat PC. Also, incorporation 4 wt% of resin content clay nanoparticles caused a significant enhancement in impact strength of studied PC by about 260% [18]. Izod impact resistances of both the polyester and the epoxy concrete were also superior to that of the cement concrete. A comparative study showed that epoxy concrete had a greater impact strength than polyester concrete [39]. Impact energy of lightweight epoxy PC improves by increasing resin content from 10 to 16 wt%, as illustrated in Figure 6.4 [40].

The drop tower impact method is another implemented procedure for evaluating the impact resistance of PCs. As illustrated in Figure 6.5, addition of MWCNTs remarkably improves the residual resistance of epoxy PC due to the vibration dissipation ability of MWCNTs [41]. The improvement may be attributed to the chemical interaction of carboxylic groups of MWCNTs and epoxy groups in the contact areas.

6.5 FRACTURE TOUGHNESS

Fracture toughness (K_{Ic}), a material resistance to crack propagation, is the main parameter that predicts toughness as the crack propagates. Another main parameter is fracture energy, G_f (the required energy to form a unit crack surface), determined to predict cumulative energy during crack propagation. G_f is equal to the area defined by softening law [42]. The fracture problem of the PCs is connected with the process of formation and development of cracks [43].

The PCs have higher fracture toughness than OPCs (up to 4.5 times) [44–47]. Although PCs exhibited higher resistance to crack opening than OPC concretes, due to brittle failure manner, enhancing their post-peak stress–strain behavior is crucial [48].

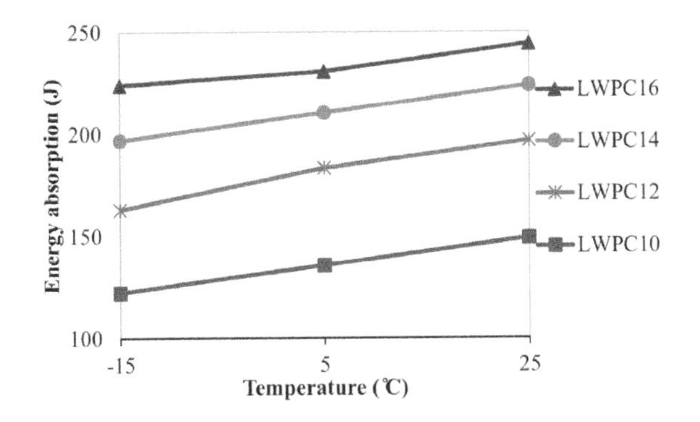

FIGURE 6.4 Impact energy of epoxy PC with 10–16 wt% epoxy content [40]

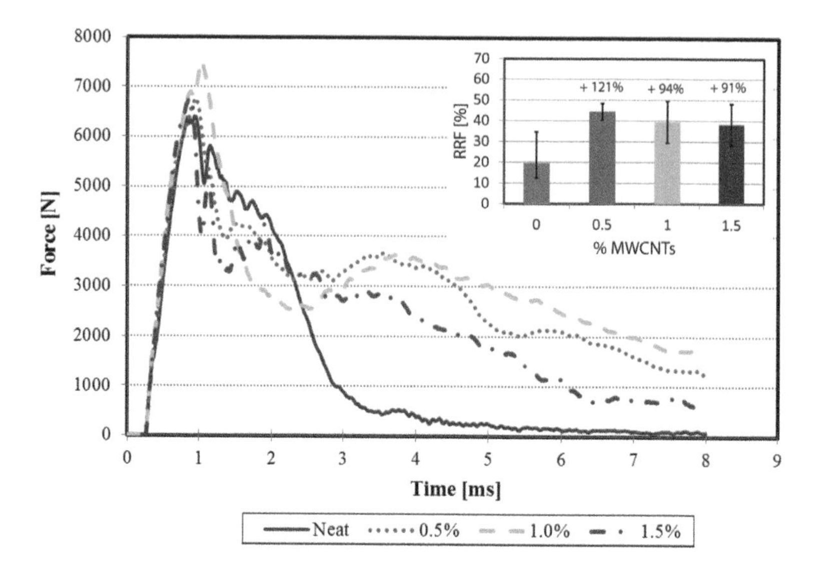

FIGURE 6.5 Variation of impact loads for epoxy PCs with different MWCNTs contents [41]

The direct uniaxial tensile test is the ideal procedure to assess the fracture energy, G_f of materials. However, it cannot be easily performed on concrete specimens. Therefore, three-point flexural tests according to RILEM, TC50-FMC [49,50], ACI 446 (ACI 2009a) [28], ASTM-D 5045-99 [20], and so on were implemented in literature as alternative methods to perform fracture tests at room temperature.

Aliha et al. [29] evaluated the fracture toughness (K_{Ic}) and indirect tensile (σ_t) of an epoxy-based PC. They affirmed the applicability of the center-cracked Brazilian disc (CBD) samples, exposed to diametric compressive load for this purpose (Figure 6.6a). Aliha et al. [33] proved the appropriateness of edge cracked (Figure 6.6b) and uncracked SCB samples for evaluating the fracture toughness (K_{Ic}) and tensile strength (σ_t) of glass fiber-reinforced PC, respectively. Also, their comparison study revealed a linear relationship between the fracture toughness and fracture energy of the PCs [51].

Critical crack tip opening displacement of the original pre-crack tip, CTODs, is obtained using the maximum applied load and the magnitude of the influential critical crack length, which is the initial notch depth added to the stable crack growth at peak load [50]. To identify fracture toughness and CTOD, which both are size-independent parameters, the Two Parameter Method (TPM) [53] can be implemented. The TPM is a direct method to evaluate fracture properties, using three-point bending under quasi-static loading. Also, for determining the fracture energy of PC, the RILEM TC 50-FMC recommendation was used [54]. The dissipated G_f can be evaluated by determining the area under the load–CMOD curves. Also, post-cracking carrying potential can be observed using load–CMOD graphs [28].

Ghasemi-ghalebahman et al. [52] experimentally determined the effect of the SCB sample size and crack angle on the fracture toughness performances of PC

FIGURE 6.6 Fracture toughness tests: (a) Compression test on cracked BD specimen [29] and (b) three-point bending test on cracked SCB specimen [52]

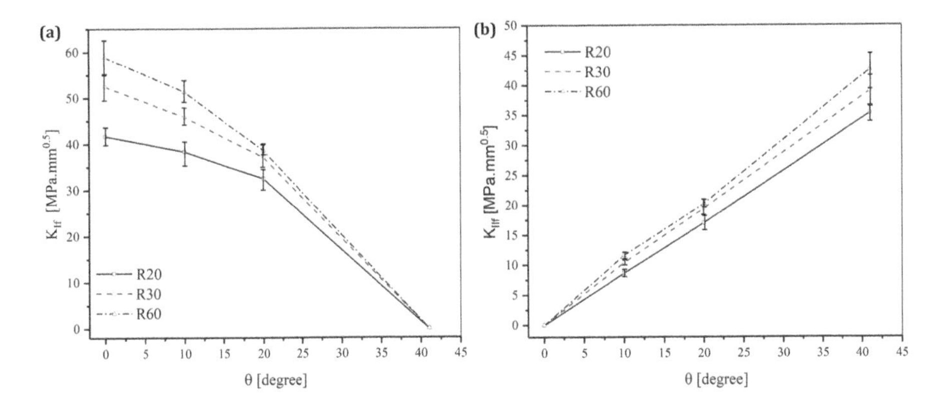

FIGURE 6.7 The behavior of (a) K_{If} and (b) K_{IIf} in terms of different crack angles for three sample sizes [52]

reinforced with clay nanoparticles. Results proved that the maximum K_{If} and the minimum K_{IIf} are obtained in the crack angle of 0° (see Figure 6.7). By increasing the crack angle from 0° up to 41°, K_{IIf} enhanced from the minimum value up to the maximum value. By increasing the SCB specimen radius to three times the primary radius, the K_{If} and K_{IIf} increased by 41.1% and 20.7%, respectively.

The fracture energy, G_f, is a size-dependent value. It was reported that increasing of notch/depth ratio increased the G_f. PC with the lower notch/depth ratio has greater strength. In general, the fracture energy of the PCs is lower than the OPCs [44]. The size dependence is largely due to the unrestorable failures outside the cracking plane, which incline to enhance with the sample size [49].

Critical stress intensity factor (K_{Ic}) and limit the speed of crack development (C) of the epoxy and furan PCs were determined in Ref. [43] as fracture parameters. From the experiments, the ratio between dynamic and critical stress intensity coefficients was obtained: $K_D = 2.5\ K_{Ic}$. For both resins, enhancing the epoxy amount from 6 to 10 wt%, led to an abrupt enhancement of K_{Ic} up to 40% and enhancing epoxy amount from 10 wt% to 22 wt% K_{Ic} increased up to 5%. Besides, the reduction of epoxy content from 20 wt% to 10 wt% didn't influence the running crack rate [43].

A comparative study exhibited that epoxy PC has more outstanding fracture toughness than unsaturated polyester (UP) ones. Evaluating the effect of aggregates on fracture properties presented that the fracture toughness of both epoxy and UP PC, including recycled foundry sand, was higher than the PC with fresh sand [55].

Studying of Mode-I fracture toughness (K_{Ic}) of the single edge notch bending (SENB) PC specimens exposed to three freeze/thaw cycles exhibited that thawing and heat-to-cool cycles had the most negative and positive effects on fracture toughness, respectively. According to the experimental results, K_{Ic} was decreased by raising the average temperature of the studied cycles [46].

One of the most applicable methods to enhance the fracture toughness of the PCs is to reinforce them with fibers. Experimental investigation exhibited that incorporating carbon fiber and glass fiber into epoxy-based PC improved the fracture toughness by 29% and 13%, respectively [45]. Also, the inclusion of glass fiber improved the fracture energy by 10% [56].

Matrix cracking, fiber breakage, fiber/matrix debonding, and fiber pull-out are the mechanisms that slow the crack propagation and increase the fracture energy in fiber-reinforced concretes. Implementing chopped coconut and sugar cane bagasse fibers increased the K_{Ic} and G_f of PC. The banana pseudostem fiber only increased the G_f [57].

In literature [48], K_{Ic} and CTOD of the fiber-reinforced PC were studied. Increasing carbon and nylon fiber and also resin content increased the critical K_{Ic} and the critical CTOD values. Incorporating the carbon fiber increased K_{Ic} and CTOD by 86.5% and 321.7%, respectively. Also, implementing the nylon fiber increased K_{Ic} and CTOD by 49.1% and 219.5%, respectively. It was reported that carbon fiber reinforcement provided better fracture properties than nylon fiber reinforcement. Fibers produce a bridging effect at the crack tip that prevents crack propagation. So, fiber-reinforced PC has a more stable crack pattern than unreinforced one [48].

Experimental assessment on the effect of fresh and used sand on the fracture features of unsaturated polyester PC showed that the substitution of fresh with recycled sand contributes to decreasing crack propagation. Furthermore, it was reported that the insertions of textile fibers from the garment industry in the PC diminished the fracture toughness, becoming the PC less resistant to crack propagation and also changed the post-peak status [42].

The effect of various nanofillers on fracture toughness properties of PC has been investigated up to now. The inclusion of both Al_2O_3 and Fe_2O_3 nanofillers into the epoxy-based PC increased the K_{Ic} and G_f because of the crack bridging and deviation. 3.0 wt% nanomaterials exhibited the best results as a toughening mechanism [50]. Incorporating nanoalumina (ANPs) improved K_{Ic} by 118%. Introducing 0.5, 1.0, 2.0, and 3.0 wt% ANPs enhanced total critical energy release rate TIC by 47.8%, 71.3%, 110.3%, and 131.8%, respectively. Improvement of fracture toughness is due to the ability of ANPs to affect polymer cross-linking [28]. The incorporation of 5 wt% clay nanoparticles increased the K_{Ic} of the PC by 7.6% [20].

6.6 BOND AND INTERFACIAL SHEAR STRENGTH

Bond strength of the PC to the other materials is produced because of the generated physical and chemical bonds through van der Waal forces and also the resin/substrate

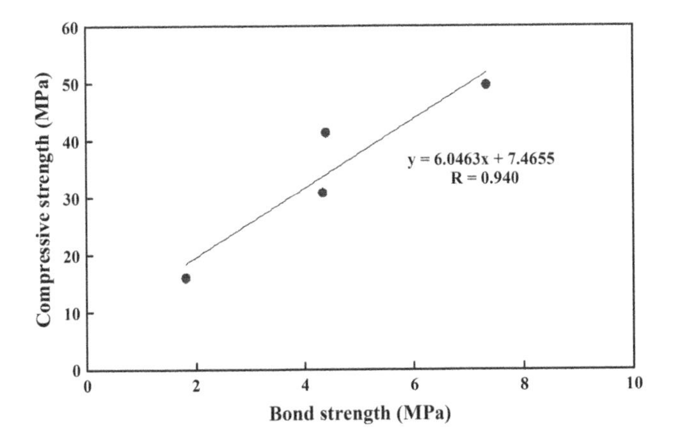

FIGURE 6.8 Compression-bonding strength relation for the polysulfide PC [3]

chemical interaction [58]. The interface properties of PC are influenced by the surface properties of materials and integrate some properties like shrinkage, rigidity, roughness, wettability, viscosity, and so on. The European Standard EN 1504 formalized repair and protection against corrosion. Surface energy, roughness, porosity, capillary suction, saturation level, and mechanical properties are the substrate parameters and properties influencing the formation and durability of the interface. On the other side, surface energy, binder setting, the kinetics of contact, thermal dilatations, shrinkage, porosity, capillarity, and mechanical properties of the PC are the important indexes that influence the interface quality [59]. Figure 6.8 demonstrates the relation between bonding strength of polysulfide PC to a steel substrate and its compressive strength [3].

Courard and Garbacz [59] studied the effect of surface parameters (surfology) of PC, especially in the case of repair efficiency. They investigated the effect of water (moisture) and demonstrated that hydrophobic treatment of substrate generally limits the loss of adhesion or cohesion of coating. From the study of Pareek et al. [60], voids may appear due to the capillary action of water at the interface. Water in bounded fresh mortar moves into capillary tubes and pores into mortar substrate, and air in the substrate migrates to the fresh mortar. Thus, air voids appear at the interface.

Bond strength tests consist of pull-off and slant shear tests. A pull-off test is performed using ASTM C1583/C1583M standard to calculate the bond strength of PC to the substrate. It is reported that the pull-off test is not the most appropriate procedure for measuring the bond strength, where tension is the dominant state of stress [61].

Sometimes the failure never happened in the pull-off test. Cohesion failure occurs inside the overlay itself, and therefore, cohesive rather than adhesive failures happen [38]. In this case, the pull-off test measured the cohesion strength of PC overlays rather than the PC/substrate adhesion bonding. Therefore, another test method, the slant shear test based on ASTM C882/C882M, is performed [1,62,63]. This method provides indirect tension force by exerting compression force that results in uniform shear stress over the bonded surface. Four types of failure modes that can be resulted from the slant shear test in PC and cement concrete bonding are [64]: Pure interfacial failure (without any crack and fracture at the substrates) (Figure 6.9a), interfacial failure mixed with minor substrate damage (Figure 6.9b), interfacial failure mixed with substrate fracture (Figure 6.9c), and substratum failure (Figure 6.9d) [1].

Similarly, three different failure modes between polyester PC and FRP bars are illustrated in Figure 6.10. Figure 6.10a shows that a part of the rod remains on the PC (wear of rod). Also, in Figure 6.10b, the failure of the rod due to tension, and in Figure 6.10c, the failure of PC due to high bonding strength can be observed [65].

PC provides high bond strength. The PC based on low viscosity vinyl ester presented a bond strength of 2.8 MPa in pull-off tests [66]. From the experimental study of Wheat and his co-workers [67], the fatigue loading of PC overlays on Portland cement concrete beams did not lead to any delamination of bond strength between both substrates up to 2 million cycles. Interface bonding strength experiments of epoxy PC overlay (including crushed basalt and crumb rubber as aggregates) to concrete substrate revealed no PC/substrate failure, and all the pull-out failures occurred inside the substrate [27]. A study of interfacial tensile bond strength of permeable PC

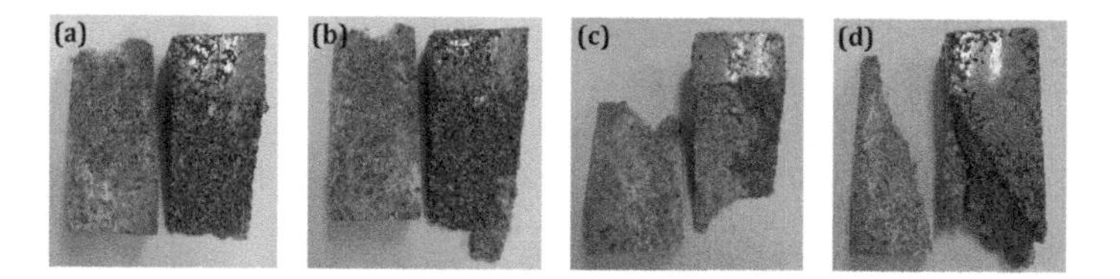

FIGURE 6.9 Slant shear test failure modes [1]

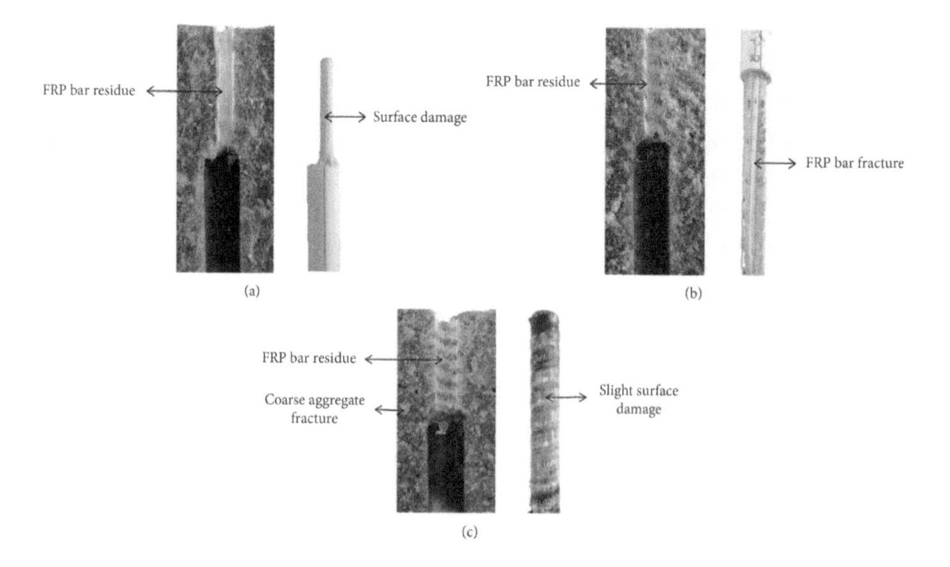

FIGURE 6.10 Three failure forms of UP/FRP bar interfacial bonding [65]

to concrete substrate indicated that curing time, interface roughness, the strength of the concrete substrate and polymer mortar, and repair position are the major effective parameters of bond behavior. It is concluded that the failure modes from the tensile bond strength are: substrate concrete failure, interlaminar failure, and mortar failure [68]. Bond strength of PC to wet substrates can be improved by applying diacrylate monomers, including zinc diacrylate (ZDA) and calcium diacrylate (CDA) [69].

Interfacial adhesion bonding could be improved by the addition of silane coupling agents. Silane coupling agents develop binder-aggregate adhesion bonding, which enhances the mechanical properties of PC. As reported, the implementation of silane coupling agent improved compressive strength and flexural strength of PC up to 15–20% higher than those of typical PC [70].

A study on the effect of implementing nanofiller on the slant shear property of PC presented that the addition of 0.5, 1.0, and 2.0 wt% MWCNTs caused a similar enhancement of 7% in apparent bond strength to the steel substrate. Inclusion of 0.5, 1.0, and 2.0 wt% nanoaluminum improved apparent bond strengths by 20, 23, and 51% above neat PC, respectively. Also, implementing 0.5 and 1 wt% silica nanoparticles increased bond strength by 7 and 18%, respectively [62].

From the pull-off test experiment, the failure mode of the PC changed from the cohesion failure mode to adhesion failure mode (between the PC and steel substrate) by incorporating MWCNTs. Therefore, MWCNTs significantly increased the cohesion strength of PC overlays [38]. Static and dynamic bond behavior under pure pull-off loads between the polyester PC and various rebars were evaluated in the literature [23]. The authors investigated the bond mechanism of both steel and GFRP rebars to PC when subjected to monotonic and cyclic tensile load. The bond strength of rebar/cementitious concrete was considerably lower than rebar/PC. Steel bars provided bonding strength 70% higher than GFRP rods.

Because of the lack of a specific standard test procedure for determining PC/steel interfacial shear strength, a new ring test procedure is implemented in the literature [71]. This procedure is based on the supposition of constant shear stress distribution between the PC and steel substrate. The interfacial shear strength can be determined by applying a simple compressive test. As aggregate dimension and resin weight content are enhanced, the interfacial shear strength is also increased. However, increasing the chopped glass fiber content reduced the interfacial shear strength. The best interfacial shear strength was obtained by incorporating the mediocre size of aggregate, the maximum resin content, and the minimum fiber content [71].

6.7 FATIGUE STRENGTH

Fatigue strength of PC remains critical mechanical properties criteria in various applications such as overlays, repairs, and machine foundations. Thermal stresses, traffic cyclic loads, degradation of the interface, and loss of flexibility in the binder are the fatigue loads that generate vertical cracks in PC overlays [67]. Due to the high ductility of PC, viscoelastic properties of polymer binder, and slow elastic strain recovery, determining the fatigue life of PC is complicated [72]. Fatigue properties of the PC are affected by various parameters like sample geometry and dimension, test procedure, loading condition, and incorporating components (resin, aggregate, …)

and their mix proportions [73]. Only limited research studied the fatigue behavior of PC. Similar to OPC concretes, a two million cycle resistance limitation has been evaluated as a stress level of 59% [74].

There is no standard fatigue test method for PC. Fatigue characteristic of PC was considered based upon S-N relationships, based upon the basic power-law functions [75]. Empirical equations of fatigue behavior of normal concrete can be implemented to predict fatigue behavior of PC as expressed in the following equation [76]:

$$L = (10)^{-aS^b(\log N)^c} \tag{6.5}$$

where L is the probability of survival, S is the stress level, N is the number of cycles to failure, and a, b, and c are experimental constants [77]. Flexural fatigue study of UP-MMA PC showed that applying MMA monomer can cause a significant decrease in the fatigue life, fatigue strength, and static strength. The author used a single-log fatigue equation (S-N relationship) to fit the fatigue life data with a 10% failure probability. The influence of MMA wt% on curve fitting parameters is shown in Figure 6.11. The correlation coefficients were obtained 0.98–0.99, regardless of MMA amount. The fatigue life of the PC illustrated narrower distributions even across the

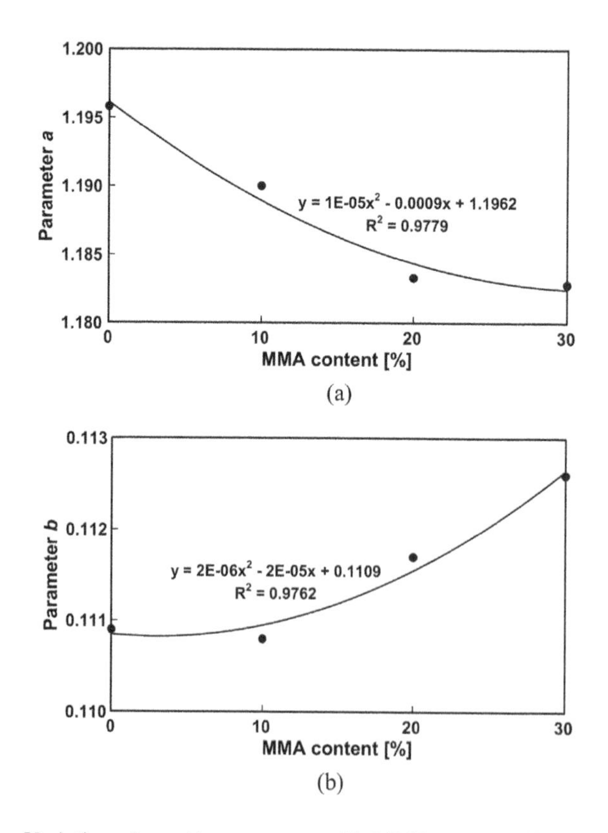

FIGURE 6.11 Variation of a and b parameters with MMA content [73]

various MMA amount. So the UP-MMA PC exhibits a dominant performance over conventional concrete in terms of fatigue stability and predictive capabilities [73].

For evaluating the flexural fatigue strength of MMA PC, beam samples were loaded at a constant rate of five cycles per second up to 2 million cycles [78]. Applying a more comprehensive range of flexural stress significantly decreased the fatigue life. Also, the fatigue life of the PC decreased when the applied stress increased. An increase in beam deflection was observed as the fatigue life was approached despite brittle failure mode. A comparative investigation revealed that the fatigue strength of the PC beam is higher than OPC concrete [78].

The authors conducted the fatigue strength of asphalt overlays using a modified four-point bending setup borrowed from AASHTO T321-07 [72]. The addition of 2.0 wt% P-MWCNTs presented an improvement in fatigue life by 55% due to ductility and fracture toughness improvement, while ANPs at the same content exhibited a decrease by 50% in fatigue life [72]. As illustrated in Figure 6.12, adding MWCNTs significantly improves the fatigue strength (failure cycles) of epoxy PC [4].

6.8 CREEP BEHAVIOR

Prediction of the time-dependent manner and persistence of PC structure is probably the foremost uncertain design aspect [79]. PC demonstrates a much more intricate time-dependent manner than OPC concrete [78]. As reported in the literature, the strength and stiffness of the PC vary with time and are intensely affected by temperature [80,81]. Unlike the OPC concrete, whose creep behavior is sensitive to humidity variations, the creep property of PC is susceptible to temperature changes [82]. Therefore, it's necessary to contemplate the impact of creep and, the creep deformation, which is larger than the elastic deformation of the PC [83]. Sometimes creep compliance (J) which is calculated by dividing the shear strain by shear stress is used to determine the creep behavior of PC.

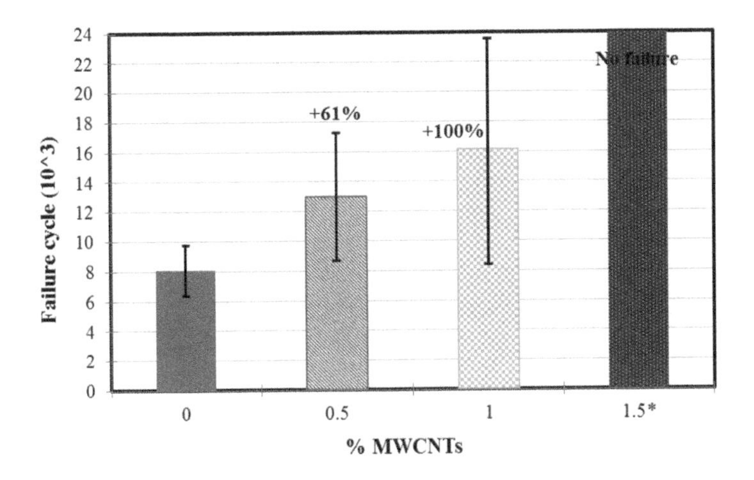

FIGURE 6.12 The behavior of fatigue strength of epoxy PC with different MWCNT contents [38]

Ribeiro and Ferreira [84] presented a new test set up for the four-point flexural creep manner of epoxy PC and regarded two viscoelastic models to suit the experimental results. The primary model is well-known empirical power law [85], and the second relies on the composition of two classical linear viscoelastic models, the Kelvin and Bruger models [86]. The authors believed that a power-law type model would be more suitable for that purpose.

Aniskevich and Hristova [87] introduced a structural approach that will be employed to evaluate the creep compliance (*J*) curves of the composites under compression either within the conditionally initial state or after ageing. It had been concluded that the results of the calculation by Kerner's model and the generalized Christensen's model matched each other and gave a decent description of the experiment. Also, the calculated values of the elastic moduli of the low-molecular polyester oligomer PC were beyond that of the experimental values. They also indicated that the *J* value of PC with different ages might be calculated from the *J* value of the appropriate binder with the structural Kerner's model [88]. The principle of analogies on ageing time and temperature is acceptable for estimating the *J* value within the range of short times. It is necessary to require under consideration both the deceleration of relaxation processes and also the shrinkage of the binder due to its physical ageing for estimation of the *J* of a PC in the range of enormous times [88].

In Ref. [89], the short-term (24 hours) uniaxial creep tests at 20°C, 30°C, and 40°C were applied on unsaturated polyester PC to obtain a Prony series equation that predicts the long-term *J* value at 20°C. The error between the predicted (proposed model) and experimental long-time creep behavior was less than 4% (see Figure 6.13).

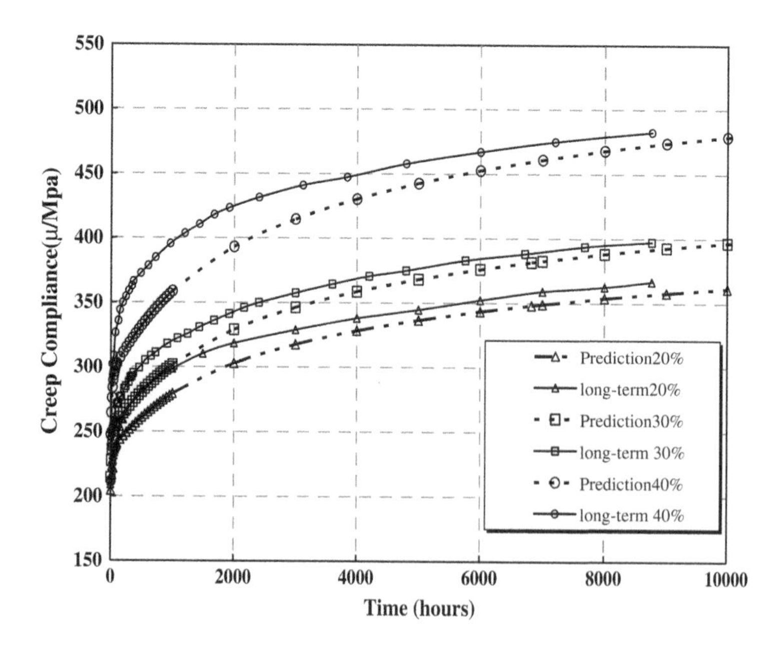

FIGURE 6.13 Variation of creep compliance on the 20%, 30%, and 40% stress ratios [89]

Due to the nonlinear viscoelastic behaviors of recycled-PET PC, the creep did not proportional to the stress ratio and increased by a rise in the stress/strength ratio.

In addition to age, the types and weight fraction of fillers and also stress/strength ratios are the foremost important factors in PC creep compliance. The creep strain of PC grows faster than OPC concrete at young ages since the creep occurs in PC as a consequence of molecular movement within the viscoelastic polymer matrix. From the experimental study, after the three-month compressive creep stress, the Poisson's ratio of PC increases by about 25% [82]. Another experimental study presents that enormous creep deformation and disastrous failure often take place when the creep stress intensity ratio is beyond about 50% [78]. At least 20% of long-term creep of PC occurs within the primary 2 days, and about 50% in the primary 20 days. Incorporating CaCO3 as a micro filler decreased the creep strain of PC significantly [89]. Also, another research on PC revealed that beyond 20% of the ultimate creep happens on the first day, and quite 90% of the ultimate creep happens within six days [90].

It was observed that the creep in MMA-based PC is about one to two times more than that of OPC concrete [78]. By increasing the stress/strength ratio, the creep increased with a nonlinear relationship. Like the research of Ref. [89], quite 20% of the ultimate creep occurred within the first day, and about 50% took place during the primary five days.

The four-point bending creep curve can be divided into three stages. In the first stage, called the unstable or transition creep stage, the creep rate continuously decreases. In the second stage, the steady creep stage, the creep variable-speed rate reaches the minimum values, which is the most research value section. In the third stage, known as the damage stage, creep speed and creep deformation increase rapidly until the material is damaged. The creep curve is not linear and shows a nonlinear viscoelastic property. When the stress level increases, the elastic strain ε_0 increases in proportion, but creep strain is greater proportion growth [83].

In Ref. [38], creep compliance of PC incorporating MWCNTs was evaluated at each creep-time interval and divided into three distinct stages as depicted in Figure 6.14. The primary stage was determined by a logarithmic equation with a high slope. The second one is a steady-state creep with a comparatively low slope and frequently covers most of the creep time of the sample. The third one is the tertiary creep with a great slope which causes the failure of the sample. It was reported that creep failure of PC occurs in an exceedingly short time, and applying higher binder content outcomes faster creep failure. However, PC to not be appropriate for applications under axial creep loading. Creep failure of PC occurs in a very similar model to compressive failure.

The creep behavior of two types of polyester PCs with various volume fractions of diabase under compression load at different stress levels was evaluated. A direct correlation between the interphase volume fraction and the instantaneous compliance of the PC was obtained [91]. The presented model is a nonlinear phenomenological model and is appropriate for the outline of the experiments on stress relaxation and creep at repeated loading. As reported, the stress dependence of creep compliance is more pronounced for a PC with a softer binder. The distinction in the region of the nonlinearity of the relative creep compliance of the two polymers at high

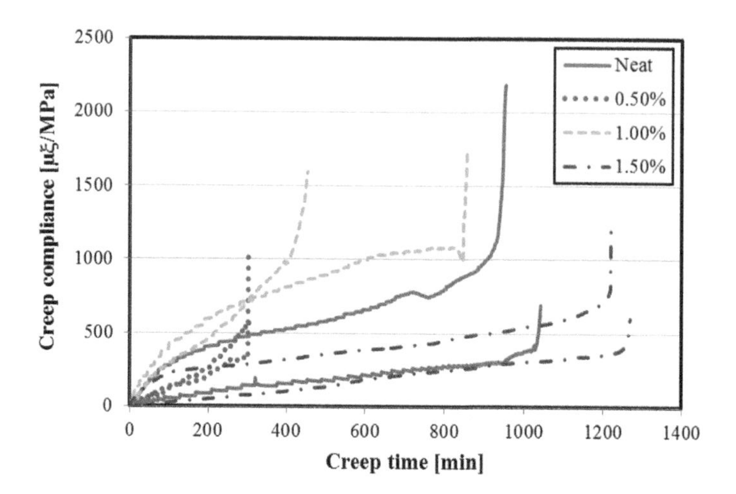

FIGURE 6.14 Creep behavior of PC incorporating different wt% of MWCNTs [38]

compression stress levels can be attributed to the difference in their porosity [91]. The long-term creep strain behavior of polyester-based PC under uniaxial compression, which presented a logarithmic function, seems to continue after five years [92].

6.9 VIBRATION AND DAMPING PROPERTIES

Due to some PC applications, including machine bed components, it is important to study the vibration properties and damping characteristics of these materials. It can be concluded from the research that has been conducted on the vibration analysis of PCs that they have decent damping characteristics [70] and are used in the manufacturing of machine tool beds [93]. The damping feature of the polyester-based PC was obtained four to seven times higher than cast iron.

Bai et al. [94] investigated the influences of the component proportion on damping characteristics of the glass-fiber-reinforced PC. Granite aggregate proportion and glass fiber dosage are the most important and the least effective factors in the damping ratio of the studied PC, respectively. Incorporating more resin enhances the damping ratio of PC (see Figure 6.15) due to the slippage trend of molecular chains in the epoxy binder, which smartly changes the vibration energy to heat energy.

Chod et al. [95] considered the possibility of fabricating steel–PC frames for increasing lathe machining stability. In the prepared steel–PC composite frames, a significant reduction of tool-workpiece FRF amplitudes was reported compared to the steel variant.

A comparative study of the damping characteristic of a reinforced PC and U-shaped steel profile revealed that the PC decreased vibration acceleration deviation up to 93.5% just in 5 ms, about three times faster than the steel profile as demonstrated in Figure 6.16 [96].

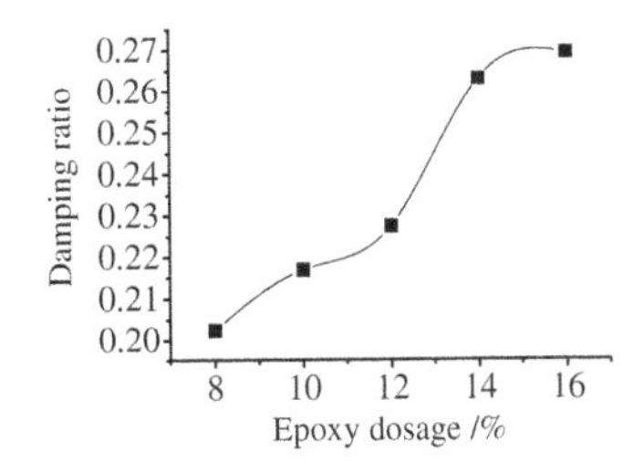

FIGURE 6.15 Variation of damping ratio of epoxy-based PC in terms of epoxy content [94]

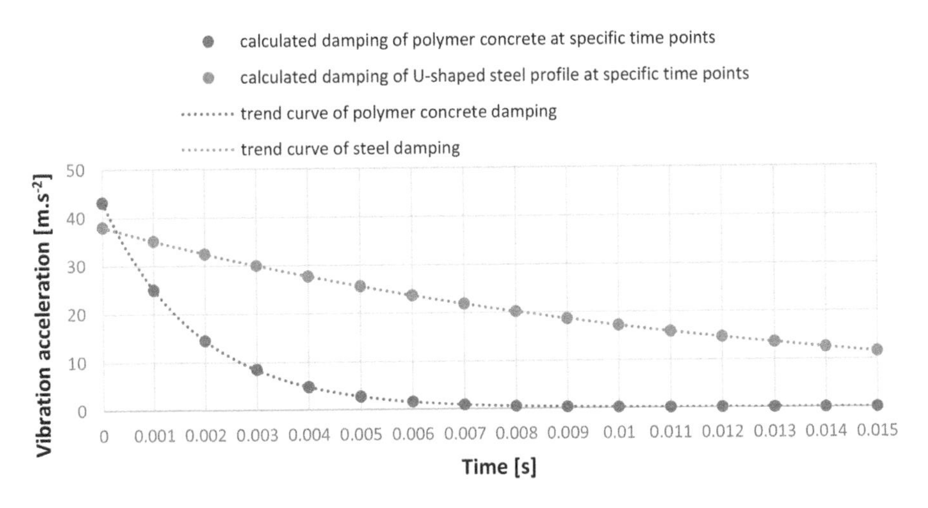

FIGURE 6.16 Comparative study of damping properties of PC and steel profile [96]

From the results of the experimental and numerical signal processing analysis, an epoxy-based PC as a machine bed component considerably limited vibrations arising and can be implemented as a viable solution for vibration damping [97]. Exposure of PC to high temperatures leads to changes in vibrational properties. Natural frequencies of the PC decreased by rising the temperature from 25°C up to 250°C [98]. Kepczak [99] evaluated the influences of adding styrene-butadiene rubber on the dynamic behavior of PC via modal testing. Introducing 10% styrene-butadiene significantly increased the dumping ratio by 431.5%.

Kwon et al. [100] introduced concrete structures consisting of cementitious and PC for the improvement of dynamic characteristics. The results demonstrated that implementing only a small volume fraction of PC in the presented complex composite caused a large vibration reduction.

6.10 ABRASION AND WEAR RESISTANCE

The abrasion and skid resistance of PC are important properties, especially in some applications such as bridge-decks, airport runways, motor-highway surfaces, and floor constructions [101]. Mani et al. [39] performed an abrasion test on Dorry's abrasion tester to evaluate the coefficient of friction of epoxy and polyester PC. The skid resistance tests were conducted on the polished surfaces of the specimens, regarding the test procedure described in BS: 812 Part III, 197516. The obtained coefficient of friction for both studied PCs are higher than that of the OPC. Besides, the results exhibited that the epoxy PC provides slightly higher abrasion resistance than the polyester PC.

In the literature [102], underwater abrasion resistance was tested on the cylindrical rubber-based PC specimens in keeping with ASTM C 1138 standard. Results presented that the PC had high underwater abrasion resistance characteristics. The abrasion effect was simulated by circulating water containing small grindstone balls over the section of the PC, which is located in water. The obtained mass change and the average depth of abrasion after 72 hours of exposure were negligible. In the PC with higher elastic deformation capacity (specimens with higher compressive strength to elastic modulus ratio), when solid water-borne particles generated abrasion loading to the submerged surface, faced to elastic deformation of PC and therefore, loss of concrete mass is smaller.

REFERENCES

[1] I. Roh, K. Jung, S. Chang, Y. Cho, Characterization of compliant polymer concretes for rapid repair of runways, *Constr. Build. Mater.* 78 (2015) 77–84. https://doi.org/10.1016/j.conbuildmat.2014.12.121.

[2] M.R. Mohamed, S.N. Aldeen, R. Abdulrazza, A study of compression strength and flexural strength for polymer concrete, *Iraqi J. Sci.* 57 (2016) 2677–2684.

[3] S. Hong, Influence of curing conditions on the strength properties of polysulfide polymer concrete, *Appl. Sci.* 7 (2017) 833. https://doi.org/10.3390/app7080833.

[4] V. Toufigh, D. Ph, S.M. Shirkhorshidi, M. Hosseinali, Experimental investigation and constitutive modeling of polymer concrete and sand interface, *Int. J. Geomech.* 17 (2016) 1–11. https://doi.org/10.1061/(ASCE)GM.1943-5622.0000695.

[5] M. Hassani, N. Abdolhosein, F. Morteza, G. Ahangari, Mechanical properties of epoxy/basalt polymer concrete : Experimental and analytical study, *Struct. Concr.* 19 (2018) 366–373. https://doi.org/10.1002/suco.201700003.

[6] V. Toufigh, M. Hosseinali, S. Masoud, Experimental study and constitutive modeling of polymer concrete's behavior in compression, *Constr. Build. Mater.* 112 (2016) 183–190. https://doi.org/10.1016/j.conbuildmat.2016.02.100.

[7] W. Lokuge, T. Aravinthan, Effect of fly ash on the behaviour of polymer concrete with different types of resin, *Mater. Des.* 51 (2013) 175–181. https://doi.org/10.1016/j.matdes.2013.03.078.

[8] J.M.L. dos Reis, Effect of textile waste on the mechanical properties of polymer concrete, *Mater. Res.* 12 (2009) 63–67.

[9] H. Haddad, M. Al, Influence of moisture content on the thermal and mechanical properties and curing behavior of polymeric matrix and polymer concrete composite, *Mater. Des.* 49 (2013) 850–856. https://doi.org/10.1016/j.matdes.2013.01.075.

[10] O. Elalaoui, E. Ghorbel, V. Mignot, M. Ben Ouezdou, Mechanical and physical properties of epoxy polymer concrete after exposure to temperatures up to 250°C, *Constr. Build. Mater.* 27 (2012) 415–424. https://doi.org/10.1016/j.conbuildmat.2011.07.027.

[11] M. Jamshidi, A. Reza, Modified polyester resins as an effective binder for polymer concretes, *Mater. Struct.* 45 (2012) 521–527. https://doi.org/10.1617/s11527-011-9779-9.

[12] M. Jamshidi, A.R. Pourkhorshidi, A comparative study on physical/mechanical properties of polymer concrete and portland cement concrete, *Asian J. Civ. Eng.* 11 (2010) 421–432.

[13] V. Toufigh, V. Toufigh, H. Saadatmanesh, S. Ahmari, Strength evaluation and energy-dissipation behavior of fiber-reinforced polymer concrete, *Adv. Civ. Eng. Mater.* 2 (2013) 622–636. https://doi.org/https://doi.org/10.1520/ACEM20130074.

[14] S.-W. Son, J.H. Yeon, Mechanical properties of acrylic polymer concrete containing methacrylic acid as an additive, *Constr. Build. Mater.* 37 (2012) 669–679. https://doi.org/https://doi.org/10.1016/j.conbuildmat.2012.07.093.

[15] N.J. Jin, K.S. Yeon, S.H. Min, J. Yeon, Using the maturity method in predicting the compressive strength of vinyl ester polymer concrete at an early age, *Adv. Mater. Sci. Eng.* 2017 (2017) 1–12. https://doi.org/10.1155/2017/4546732.

[16] K. Yeon, Stress – Strain curve modeling and length effect of polymer concrete subjected to flexural compressive stress, *J. Appl. Polym. Sci.* 114 (2009) 3819–3826. https://doi.org/10.1002/app.

[17] S.R. da Silva, F.N. Cimadon, P.M. Borges, J.Z. Schiavon, E. Possan, J.J. de Oliveira Andrade, Relationship between the mechanical properties and carbonation of concretes with construction and demolition waste. *Case Stud. Constr. Mater.* 16 (2022) e00860. https://doi.org/10.1016/j.cscm.2021.e00860.

[18] M.H. Niaki, A. Fereidoon, M.G. Ahangari, Experimental study on the mechanical and thermal properties of basalt fiber and nanoclay reinforced polymer concrete, *Compos. Struct.* 191 (2018) 231–238. https://doi.org/10.1016/j.compstruct.2018.02.063.

[19] M. Golestaneh, G. Amini, G.D. Najafpour, M.A. Beygi, Evaluation of mechanical strength of epoxy polymer concrete with silica powder as filler, *World Appl. Sci.* J. 9 (2010) 216–220.

[20] M.M. Shokrieh, A.R. Kefayati, M. Chitsazzadeh, Fabrication and mechanical properties of clay/epoxy nanocomposite and its polymer concrete, *Mater. Des.* 40 (2012) 443–452. https://doi.org/10.1016/j.matdes.2012.03.008.

[21] J.P. Gorninski, D.C. Dal Molin, C.S. Kazmierczak, Strength degradation of polymer concrete in acidic environments, *Cem. Concr. Compos.* 29 (2007) 637–645. https://doi.org/10.1016/j.cemconcomp.2007.04.001.

[22] A.J.M. Ferreira, A.T. Marques, M.C.S. Ribeiro, P.R. N, Flexural performance of polyester and epoxy polymer mortars under severe thermal conditions, *Cem. Concr. Compos.* 26 (2004) 803–809.

[23] J. Toma, Reinforced polymer concrete: Physical properties of the matrix and static/dynamic bond behaviour, *Cem. Concr. Compos.* 27 (2005) 934–944. https://doi.org/10.1016/j.cemconcomp.2005.06.004.

[24] H. Haddad, *Optimisation of Polymer Concrete for the Manufacture of the Precision Tool Machines Bases*, Swinburne University of Technology, 2013.

[25] A.J.M.F.M.C.S. Riberio, C.M.C. Pereira, S.P.B. Sousa, P.R.O. Novoa, Fire reaction and mechanical performance analyses of polymer concrete materials modified with micro and nano alumina particles, *Restor. Build. Monum.* 19 (2013) 195–202.

[26] W. Lokuge, T. Aravinthan, Effect of fly ash on the behaviour of polymer concrete with different types of resin, *Mater. Des.* 51 (2013) 175–181. https://doi.org/10.1016/j.matdes.2013.03.078.

[27] J. Wang, Q. Dai, S. Guo, R. Si, Mechanical and durability performance evaluation of crumb rubber-modified epoxy polymer concrete overlays, *Constr. Build. Mater.* 203 (2019) 469–480. https://doi.org/10.1016/j.conbuildmat.2019.01.085.

[28] M. Emiroglu, A.E. Douba, R.A. Tarefder, U.F. Kandil, M.R. Taha, New polymer concrete with superior ductility and fracture toughness using alumina nanoparticles, *J. Mater. Civ. Eng.* 29 (2017) 1–9. https://doi.org/10.1061/(ASCE)MT.1943-5533.0001894.

[29] M.R.M. Aliha, M.M. Shokrieh, M.R. Ayatollahi, Determination of tensile strength and crack growth of a typical polymer concrete using circular disc samples, *J. Mech. Res. Appl.* 4 (2012) 49–56.

[30] W. Lokuge, T. Aravinthan, Mechanical properties of polymer concrete with different types of resin, in: *22nd Australas. Conf. Mech. Struct. Mater.*, Sydney, Australia, 2012. https://doi.org/10.1201/b15320-204.

[31] G. Matinez Barrera, E. Vigueras Santiago, O. Gencel, H.E. Hagg Lobland, Polymer concretes: A description and methods for modification and improvement, *Mater. Educ.* 33 (2011) 37–52.

[32] B.-W. Jo, S.-K. Park, D.-K. Kim, Mechanical properties of nano-MMT reinforced polymer composite and polymer concrete, *Constr. Build. Mater.* 22 (2008) 14–20. https://doi.org/10.1016/j.conbuildmat.2007.02.009.

[33] M.R.M. Aliha, M. Heidari-Rarani, M.M. Shokrieh, M.R. Ayatollahi, Experimental determination of tensile strength and KIc of polymer concretes using semi-circular bend (SCB) specimens, *Struct. Eng. Mech.* 43 (2012) 823–833. https://doi.org/10.12989/sem.2012.43.6.823.

[34] M.H.W. Ibrahim, A.F. Hamzah, N. Jamaluddin, P.J. Ramadhansyah, A.M. Fadzil, Split tensile strength on self-compacting concrete containing coal bottom ash, *Procedia - Soc. Behav. Sci.* 195 (2015) 2280–2289. https://doi.org/10.1016/j.sbspro.2015.06.317.

[35] N.N. Gerges, C.A. Issa, S. Fawaz, Effect of construction joints on the splitting tensile strength of concrete, *Case Stud. Constr. Mater.* 3 (2015) 83–91. https://doi.org/10.1016/j.cscm.2015.07.001.

[36] M. Hassani, N. Abdolhosein, F. Morteza, G. Ahangari, Mechanical properties of epoxy / basalt polymer concrete : Experimental and analytical study, *Struct. Concr.* 19 (2018) 366–373. https://doi.org/10.1002/suco.201700003.

[37] M. Bărbuţă, M. Harja, I. Baran, Comparison of mechanical properties for polymer concrete with different types of filler, *J. Mater. Civ. Eng.* 22 (2010) 696–701. https://doi.org/10.1061/(ASCE)MT.1943-5533.0000069.

[38] S.M. Daghash, *New Generation Polymer Concrete Incorporating Carbon Nanotubes*, The University of New Mexico, Thesis for Master of Science, 2013.

[39] P. Mani, A.K. Gupta, S. Krishnamoorthy, Comparative study of epoxy and polyester resin-based polymer concretes, *Int. J. Adhes. Adhes.* 7 (1987) 157–163. https://doi.org/10.1016/0143-7496(87)90071-6.

[40] F. Heidarnezhad, K. Jafari, T. Ozbakkaloglu, Effect of polymer content and temperature on mechanical properties of lightweight polymer concrete, *Constr. Build. Mater.* 260 (2020) 119853. https://doi.org/10.1016/j.conbuildmat.2020.119853.

[41] S.M. Daghash, E.M. Soliman, U.F. Kandil, M.M. Reda Taha, Improving impact resistance of polymer concrete using CNTs, *Int. J. Concr. Struct. Mater.* 10 (2016) 539–553. https://doi.org/10.1007/s40069-016-0165-4.

[42] M. Antonio, G. Jurumenha, J. Marciano, Fracture mechanics of polymer mortar made with recycled raw materials, *Mater. Res.* 13 (2010) 475–478.

[43] M.T. Hamza, Research of polymer concrete fracture parameters, *Earth Environ. Sci.* (2017) 1–4.

[44] J.M.L. Reis, A.J.M. Ferreira, The influence of notch depth on the fracture mechanics properties of polymer concrete, *Int. J. Fract.* 124 (2003) 33–42.

[45] J.M. Reis, A.J. Ferreira, Assessment of fracture properties of epoxy polymer concrete reinforced with short carbon and glass fibers, *Constr. Build. Mater.* 18 (2004) 523–528. https://doi.org/10.1016/j.conbuildmat.2004.04.010.

[46] M. Heidari-Rarani, M.R.M. Aliha, M.M. Shokrieh, M.R. Ayatollahi, Mechanical durability of an optimized polymer concrete under various thermal cyclic loadings – An experimental study, *Constr. Build. Mater.* 64 (2014) 308–315. https://doi.org/10.1016/j.conbuildmat.2014.04.031.

[47] Y. Ohama, *Handbook of Polymer-Modified Concrete and Mortars*, 1st ed., William Andrew Publishing, Koriyama, Japan, 1995. https://doi.org/https://doi.org/10.1016/B978-081551358-2.50001-3.
[48] S. Park, B. Jo, J. Park, J. Choi, Fracture behaviour of polymer concrete reinforced with carbon and nylon fibres, *Adv. Cem. Res.* 22 (2010) 45–51. https://doi.org/10.1680/adcr.2008.22.1.45.
[49] J.M.L. Reis, A.J.M. Ferreira, A contribution to the study of the fracture energy of polymer concrete and fibre reinforced polymer concrete, *Polym. Test.* 23 (2004) 437–440. https://doi.org/10.1016/j.polymertesting.2003.09.008.
[50] J.M.L. Reis, D.C. Moreira, L.C.S. Nunes, L.A. Sphaier, Evaluation of the fracture properties of polymer mortars reinforced with nanoparticles, *Compos. Struct.* 93 (2011) 3002–3005. https://doi.org/10.1016/j.compstruct.2011.05.002.
[51] M. Asdollah-Tabar, M. Heidari-Rarani, M.R.M. Aliha, The effect of recycled PET bottles on the fracture toughness of polymer concrete, *Compos. Commun.* 25 (2021) 100684. https://doi.org/10.1016/j.coco.2021.100684.
[52] A. Ghasemi-ghalebahman, A.A. Aghdam, S. Pirmohammad, M.H. Niaki, Experimental investigation of fracture toughness of nanoclay reinforced polymer concrete composite: Effect of specimen size and crack angle, *Theor. Appl. Fract. Mech.* 117 (2022).
[53] S. Shah, A. Carpinteri, *Fracture Mechanics Test Methods for Concrete RILEM*, 1st ed., CRC Press, 1991.
[54] RILEM, TC 50-FMC. Fracture mechanics of concrete, determination of fracture energy of mortar and concrete by means of three-point bend test on notched beams, RILEM Recomm. *Mater. Struct.* 18 (1995) 407–413.
[55] J.M.L. dos Reis, M.A.G. Jurumenha, Experimental investigation on the effects of recycled aggregate on fracture behavior of polymer concrete, *Mater. Res.* 14 (2011) 326–330. https://doi.org/10.1590/S1516-14392011005000060.
[56] H. reza Karimi, M.R.M. Aliha, Statistical assessment on relationship between fracture parameters of plain and fiber reinforced polymer concrete materials, *Compos. Commun.* 28 (2021) 100969. https://doi.org/https://doi.org/10.1016/j.coco.2021.100969.
[57] J.M.L. Reis, Fracture and flexural characterization of natural fiber-reinforced polymer concrete, *Constr. Build. Mater.* 20 (2006) 673–678. https://doi.org/10.1016/j.conbuildmat.2005.02.008.
[58] A.V. Pocius, *Adhesion and Adhesives Technology*, 3rd ed., Hanser, Munich, 2012. https://doi.org/https://doi.org/10.3139/9783446431775.fm.
[59] L. Courard, A. Garbacz, Surfology: what does it mean for polymer concrete composites?, *Restor. Build. Monum.* 16 (2010) 291–302. http://orbi.ulg.ac.be/jspui/handle/2268/30248.
[60] S.N. Pareek, Y. Ohama, K. Demura, Adhesion mechanism of ordinary cement mortar to mortar substrates by polymer dispersion coating, in: *Int. Congr. Polym. Concr. (ICPIC)*, Shanghai, China, 1990: pp. 442–449.
[61] A. Momayez, M.R. Ehsani, A.A. Ramezanianpour, H. Rajaie, Comparison of methods for evaluating bond strength between concrete substrate and repair materials, *Cem. Concr. Res.* 35 (2005) 748–757. https://doi.org/https://doi.org/10.1016/j.cemconres.2004.05.027.
[62] A. Douba, M. Genedy, E. Matteo, U.F. Kandil, M.M.R. Taha, The significance of nanoparticles on bond strength of polymer concrete to steel, *Int. J. Adhes. Adhes.* 74 (2017) 77–85. https://doi.org/10.1016/j.ijadhadh.2017.01.001.
[63] A.G. Bajgirani, S. Moghadam, A. Arbab, H. Vatankhah, The mechanical characteristics of polymer concrete using polyester resin, *J. Fundam. Appl. Sci.* ISSN. 8 (2016) 571–578. https://doi.org/http://dx.doi.org/10.4314/jfas.v8i3s.238.
[64] B.A. Tayeh, B.H. Abu Bakar, M.A. Megat Johari, Y.L. Voo, Mechanical and permeability properties of the interface between normal concrete substrate and ultra high performance fiber concrete overlay, *Constr. Build. Mater.* 36 (2012) 538–548. https://doi.org/https://doi.org/10.1016/j.conbuildmat.2012.06.013.

[65] W. Li, M. Zhou, F. Liu, Y. Jiao, Q. Wu, Experimental study on the bond performance between fiber-reinforced polymer bar and unsaturated polyester resin concrete, *Adv. Civ. Eng.* 2021 (2021). https://doi.org/10.1155/2021/6676494.
[66] L. Czarneck, B. Chmielewska, The influence of coupling agent on the properties of vinylester mortar, in: Y.O. and M. Puterman (Ed.), *Second Int. RILEM Symp. Adhes. between Polym. Concr.*, RILEM Publications SARL, 1999: pp. 57–65.
[67] B.D.L. Wheat, D.W. Fowler, A.I. Ai-negheimish, Thermal and fatigue behavior of polymer concrete overlaid beams, *J. Mater. Civ. Eng.* 5 (1993) 460–477.
[68] H. Huang, B. Liu, K. Xi, T. Wu, Interfacial tensile bond behavior of permeable polymer mortar to concrete, *Constr. Build. Mater.* 121 (2016) 210–221. https://doi.org/10.1016/j.conbuildmat.2016.05.149.
[69] N. Ahn, Effects of diacrylate monomers on the bond strength of polymer concrete to wet substrates, *J. Appl. Polym. Sci.* 90 (2003) 991–1000.
[70] R. Bedi, R. Chandra, S.P. Singh, Mechanical properties of polymer concrete, *J. Compos.* 2013 (2013) 1–12. https://doi.org/10.1155/2013/948745.
[71] M.M. Shokrieh, M. Heidari-Rarani, M. Shakouri, E. Kashizadeh, Effects of thermal cycles on mechanical properties of an optimized polymer concrete, *Constr. Build. Mater.* 25 (2011) 3540–3549. https://doi.org/10.1016/j.conbuildmat.2011.03.047.
[72] A. Douba, *Mechanical Characterization of Polymer Concrete with Nanomaterials*, The University of New Mexico, Thesis for Master of Science, 2017.
[73] K. Yeon, Y. Choi, K. Kim, J. Heum, Flexural fatigue life analysis of unsaturated polyester-methyl methacrylate polymer concrete, *Constr. Build. Mater.* 140 (2017) 336–343. https://doi.org/10.1016/j.conbuildmat.2017.02.116.
[74] K.Kobayashi, Y.Ohama, T. Ito, Fatigue properties of resin concrete under repeated compression loads, *Seisan Kenkyu.* 26 (1974) 116–118.
[75] G.A. Woelfl, M. McNerney, C.J. Chang, Flexural fatigue of polymer concrete, *Cem. Concr. Aggregates.* 3 (1981) 84–88.
[76] J. McCALL, Probability of fatigue failure of plain concrete, *J. Am. Concr. Inst. (JOURNAL ACI).* 55 (1958) 233–244.
[77] C. Vipulanandan, S. Mebarkia, Fatigue crack growth in polyester polymer concrete, *Am. Concr. Inst.* (2001) 153–168.
[78] M.H. and D.W. Fowler, Creep and fatigue of polymer concrete, *Am. Concr. Inst.* 89 (1985) 323–342. https://doi.org/10.14359/6256.
[79] Y. Khristova, K. Aniskevich, Prediction of creep of polymer concrete, *Mech. Compos. Mater.* 31 (1995) 216–219.
[80] J. Hristova, V. Valeva, J. Ivanova, Aging and filler effects on the creep model parameters of thermoset composites, *Compos. Sci. Technol.* 62 (2002) 1097–1103. https://doi.org/https://doi.org/10.1016/S0266-3538(02)00055-6.
[81] J.D. Ferry, *Viscoelastic Properties of Polymers*, 3rd ed., Wiley, New York, 1980.
[82] K.S. Rebeiz, Time-temperature properties of polymer concrete using recycled PET, *Cem. Concr. Compos.* 17 (1995) 119–124. https://doi.org/10.1016/0958-9465(94)00004-I.
[83] S. Huicun, T. Kui, H. Yanhua, Study on the creep properties of resin concrete, *Appl. Mech. Mater.* 472 (2014) 649–653. https://doi.org/10.4028/www.scientific.net/AMM.472.649.
[84] M. Ribeiro, A. Ferreira, Creep behaviour of FRP-reinforced polymer concrete, *Compos. Struct.* 57 (2002) 47–51. https://doi.org/10.1016/S0263-8223(02)00061-2.
[85] D.W. Scott, J.S. Lai, A.-H. Zureick, Creep behavior of fiber-reinforced polymeric composites: a review of the technical Literature, *J. Reinf. Plast. Compos.* 14 (1995) 588–617. https://doi.org/10.1177/073168449501400603.
[86] W.N. Findley, *Creep and Relaxation of Nonlinear Viscoelastic Materials*, 1st ed., Dover publication, New York, 1976.

[87] K. Aniskevich, J. Hristova, Creep of polyester resin filled with minerals, *J. Appl. Polym. Sci.* 77 (2000) 45–52.

[88] K. Aniskevich, J. Hristova, J. Jansons, Physical aging of polymer concrete during creep, *J. Appl. Polym. Sci.* 89 (2003) 3427–3431.

[89] B.-W. Jo, G.-H. Tae, C.-H. Kim, Uniaxial creep behavior and prediction of recycled-PET polymer concrete, *Constr. Build. Mater.* 21 (2007) 1552–1559. https://doi.org/http://dx.doi.org/10.1016/j.conbuildmat.2005.10.003.

[90] J. Hristova, R.A. Bares, Relation between creep and performance of PC, in: *Int. Congr. Polym. Concr. (ICPIC)*, Brighton, UK, 1987: pp. 99–102.

[91] K. Aniskevich, J. Khristova, J. Jansons, Creep of polymer concrete in the nonlinear region, *Mech. Compos. Mater.* 36 (2000) 85–96.

[92] R.D. Maksimov, L.A. Jirgens, E.Z. Plume, J.O. Jansons, Water resistance of polyester polymer concrete, *Mech. Compos. Mater.* 39 (2003) 99–110. https://doi.org/10.1023/A:1023407910034.

[93] S. Orak, Investigation of vibration damping on polymer concrete with polyester resin, *Cem. Concr. Res.* 30 (2000) 171–174. https://doi.org/10.1016/S0008-8846(99)00225-2.

[94] W. Bai, J. Zhang, P. Yan, X. Wang, Study on vibration alleviating properties of glass fiber reinforced polymer concrete through orthogonal tests, *Mater. Des.* 30 (2009) 1417–1421. https://doi.org/https://doi.org/10.1016/j.matdes.2008.06.028.

[95] M. Chod, P. Dunaj, B. Powa, S. Berczy, T. Okulik, Increasing lathe machining stability by using a composite steel – polymer concrete frame, *CIRP J. Manuf. Sci. Technol.* 31 (2020) 1–13. https://doi.org/10.1016/j.cirpj.2020.09.009.

[96] O. Petruška, J. Zajac, V. Molnár, G. Fedorko, J. Tkáčˇ, The effect of the carbon fiber content on the flexural strength of polymer concrete testing samples and the comparison of polymer concrete and U-shaped steel profile damping, *Materials (Basel).* 12 (2019) 1917-. https://doi.org/10.3390/ma12121917.

[97] M. Troncossi, G. Canella, N. Vincenzi, Identification of polymer concrete damping properties, *J. Mech. Eng. Sci.* (2020) 1–10. https://doi.org/10.1177/0954406220949587.

[98] B. Imane, L. Boudjemaa, B. Ibtissam, Mechanical characterization of resin concrete subjected to high temperatures by vibration analysis, *Mater. Sci. Forum.* 962 (2019) 236–241. https://doi.org/10.4028/www.scientific.net/MSF.962.236.

[99] N. Kepczak, Influence of the addition of styrene- butadiene rubber on the dynamic properties of polymer concrete for machine tool applications, *Adv. Mech. Eng.* 11 (2019) 1–11. https://doi.org/10.1177/1687814019865841.

[100] S. Kwon, S. Ahn, H. Koh, J. Park, Polymer concrete periodic meta-structure to enhance damping for vibration reduction, *Compos. Struct.* 215 (2019) 385–390. https://doi.org/10.1016/j.compstruct.2019.02.022.

[101] R. Allahvirdizadeh, R. Rashetnia, A. Dousti, M. Shekarchi, Application of the polymer concrete in repair of concrete structures : A literature review, in: *4th Int. Conf. Concr. Repair*, Dresden, Germany, 2011: pp. 1–10. https://doi.org/10.13140/2.1.4893.7925.

[102] J. Sustersic, A. Zajc, I. Leskovar, V. Dobnikar, PC and LMC with granulated rubber made from waste car tires, in: J.B. de Aguiar, S. Jalali, A. Camoes, R.M. Ferreira (Eds.), *13th Int. Congr. Polym. Concr. (ICPIC)*, Funchal-Maderia, Portugal, 2010: pp. 289–297.

7 Thermal Properties of Polymer Concrete

Abstract

Polymer concrete (PC) structures and precast members may be exposed to elevated temperatures, different intense thermal cycles, and even fire hazards. The presence of polymer resin in the composition of this concrete and also the very high sensitivity of polymers to high temperatures, make it necessary to pay attention to the temperature behavior of these materials. Numerous research studies were conducted to evaluate the thermal characteristics and enhance the thermal stability of the PCs. After reviewing thermal expansion and thermal conductivity features, the main contribution of the present chapter is to study the stability of PC in elevated temperatures, thermal cycles, freeze–thaw conditions, fire, as well as hot water. Furthermore, the flammability properties of PC are discussed. Knowing the factors affecting the thermal behavior of PC, the appropriate strategy can be selected to improve the properties and thermal stability of these materials.

7.1 THERMAL EXPANSION

Sometimes polymer concretes (PCs) are implemented as a repair and coating material. In such cases, some failures are occurred due to incompatibility between the coating (PC) and the substrate. It is attributed to the inequality in coefficients of thermal expansion (CTE) coupled with the two materials [1]. Therefore, the evaluation of the thermal expansion manner of PCs is vital. The CTE of PC is evaluated by the CTE of the polymer binder and aggregates. The CTE of PCs is generally more than twice that of OPC [2].

The CTE was carried out from different standards, such as ASTM C531, ASTM E228-85, SR EN ISO 10545-8:2000, and RILEM TC/113 PC-13, and also a custom-built instrument [3].

The CTE of PCs increases with increasing temperature. Also, the sensitivity of thermal expansion of polymers to the temperature variations is more than that of hydraulic concretes. The CTE of the optimum PC, including basalt, sand, and fly ash, was obtained 14.9×10^{-6} °C^{-1}. It was observed that replacing the aggregates with the composition mentioned above, change the CTE. It was attributed to the interfacial adhesion bonding behavior of various particles [3].

The behavior of CTE for epoxy and unsaturated polyester PCs with the variation of temperature between −20°C and 60°C fits on a parabolic law. The CTE of both binder formulations is similar at the range of −15°C to 10°C. However, by raising the temperature by more than 10°C, the epoxy PC has a higher rate of CTE than the polyester one. The mean CTE values of polyester and epoxy PC were calculated 25 and 27 μm/m °C for below the room temperature, and 34 and 46 μm/m °C for above

DOI: 10.1201/9781003326311-7

the room temperature, respectively. It was founded that the incorporating of 1 wt% chopped glass fiber has no substantial influence on the CTE of epoxy PC, while the incorporating of 2 wt% chopped carbon fiber considerably decreased the CTE for temperatures higher than room temperature [4].

The homogeneity of the CTE in the PC structure causes a reduction in thermal stresses. The features of the aggregates do not alter during the curing process of PC. It was noted that the higher CTE of resin compared to the aggregates causes to increase in the CTE of PC in interfacial areas within the aggregates. Increasing the moisture both in resin and aggregates of PC leads to enhancing the CTE because of the generated chemical interaction between the water and the polymer binder [5].

7.2 THERMAL CONDUCTIVITY

Generally, the thermal conductivities of the different PCs are very low. The thermal conductivity (k or λ) of the epoxy PC and epoxy polyurethane acryl PC was reported 2.044 [6] and 0.425 W/m/K [7], respectively.

Different methods were implemented by authors for measuring the thermal conductivity of the PC such as STAS 5912-89 standard [7] and a hot disk system [6,8]. It was founded that decreasing the polymer to aggregate ratio enhances the porosity of PC. Therefore, the heat flow transmitted through the PC weakens, and consequently, the value of λ declines. Thus, the thermal conductivity of the PC is expressed as a function of the porosity. Increasing the porosity decreases the thermal conductivity of the PC [8].

7.3 ELEVATED TEMPERATURE RESISTANCE

Elevated temperatures have a substantial effect on the performance of PC due to the viscoelastic characteristics and thermal sensitivity of the polymer matrix [2,9–11]. Thus, the study of the thermal behavior and thermal endurance of PC seems obligatory [12]. Extremely elevated temperatures can dramatically reduce the strength of the PCs due to transforming the polymer binder from a hard form into a soft form. Exposure to high temperatures also causes polymer degradation and, therefore, diminishes the PC strength. Several works were concentrated on the behavior of PC at high temperatures. Thermal resistance studies on polymeric concrete are classified into two major categories: Hot strength test and cold strength test. In the hot test, after subjecting the samples to a high temperature for a specific time, mechanical tests are performed on the hot samples. While in the cold test the specimen is exposed at an elevated temperature, and after storage at that temperature for a specific time, the sample is gradually cooled to room temperature. A strength test is performed on it. Therefore, the obtained strength is the residual strength of the PC. The PC specimens should be post-cured at a temperature of about 60–80°C before the high-temperature test, to avoid the post-curing effects, stabilize the temperature of the samples and reduce the thermal gradients [13–15].

The hot strength test on epoxy PC at different temperatures of 24°C, 50°C, 75°C, and 100°C indicated that compressive and flexural strengths as well as splitting tensile strength rapidly diminished with elevating the temperature. Furthermore, the yield displacement raised considerably from 0.5% up to about 2.5% for elevating the temperature from 24°C to 100°C. From the comparison between the barrel-shaped failure model of the cylindrical samples at 100°C and brittle failure of them in the room temperature during the compressive test (see Figure 7.1), the authors concluded that the exposure to the high temperatures changed the polymer binder behavior from the brittle to the ductile and therefore, the PC destructed in a ductile behavior at elevated temperatures.

The mentioned changes can be seen in the compressive stress–strain behavior of the epoxy PC after exposure to high temperatures (see Figure 7.2) [17]. The stress–strain diagram of the sample at room temperature (RT) after passing the peak, has a rapid downward trend and the sample has a brittle failure. While the stress–strain diagrams of the same PC at elevated temperatures, reach a peak with a much lower slope and then do not fall rapidly. Instead, it shows a lot of strain and ductile failure occurs. It is observed that with raising the temperature, the PC strain increases. As shown in Figure 7.3, the compressive strength and modulus of elasticity of the PC decreases with increasing temperature. This reduction is much greater for temperatures above 40°C. This is due to exposure to the temperatures higher than the glass transition temperature of epoxy at which the polymer changes from solid to soft state [17].

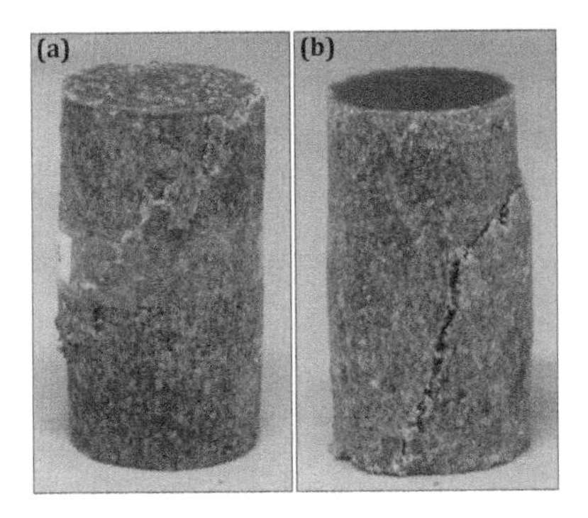

FIGURE 7.1 Failure pattern of epoxy/basalt samples exposed to (a) 24°C and (b) 100°C [16]

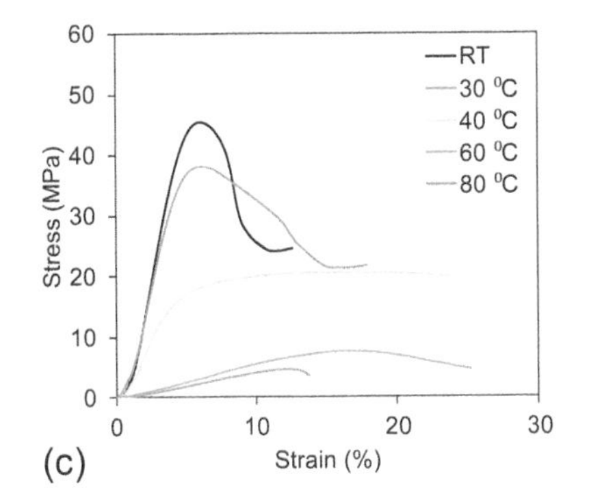

FIGURE 7.2 Compressive stress–strain behavior of the epoxy PC exposed to the high temperatures [17]

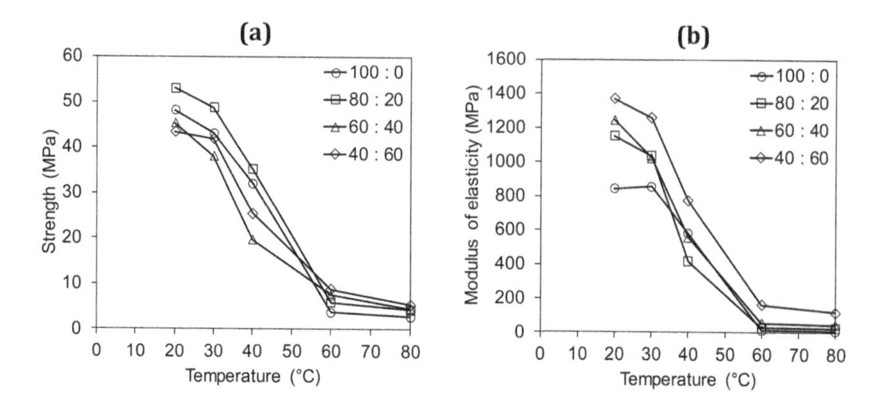

FIGURE 7.3 Behavior of the epoxy PC exposed to the high temperatures: (a) compressive strength and (b) modulus of elasticity [17]

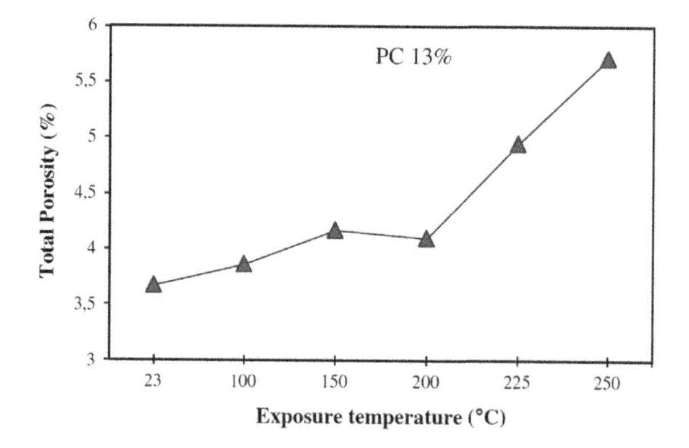

FIGURE 7.4 Effect of high temperatures on porosity of epoxy PC [8]

The resistance of the PC to heat is affected by the type of resin. For example, it is recommended that the highest tolerable temperature for the epoxy and polyester PC is about 60°C for long subjections and about 100–120°C for short subjections. The inclusion of aggregates and fillers significantly improves glass transition temperature and also flammability of the PCs compared to the neat polymer. PC illustrates better behavior than OPC in preserving its compressive strength and alleviating weight loss when exposed to a high temperature of 225°C. The elevated temperatures enhance the porosity of PC (see Figure 7.4) [8].

As represented in Figure 7.5, the thermal conductivity of PC decreased by subjecting it to elevated temperatures due to the formation of microcracks and micropores [8]. Besides, it was reported that the sensitivity of flexural and tensile performance to elevated temperature is more than the compressive strength [18].

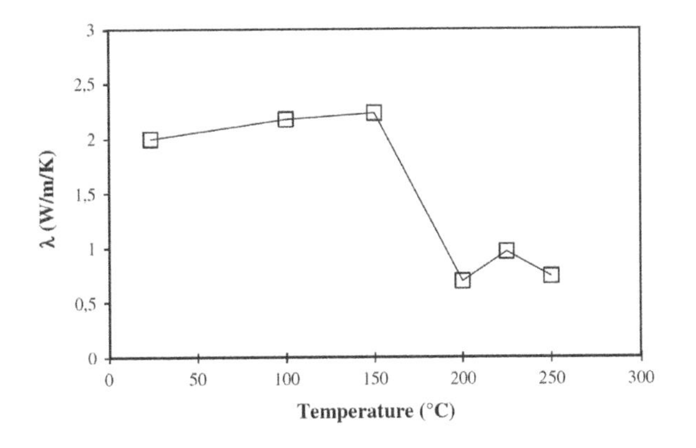

FIGURE 7.5 Effect of high temperatures on thermal conductivity of epoxy PC [8]

Blaga and Beaudoin [19] demonstrated that long-term exposure of PC to the temperature of 60°C did not influence its performance. However, considerable changes in the strength were reported for brief periods of exposure of 100–120°C.

Literature [20] considered the influence of various temperatures of 20°C, 100°C, 150°C, and 200°C on the strength of three types of PC systems fabricated by two types of epoxy and one polyester resin, which is composed of 9, 12, and 15 wt% resin. As a result of experiments, rising the temperature led to improved compressive strength and toughness. Also, only a slight increase in elastic modulus of the PCs after subjecting to elevated temperatures except grouting epoxy was founded. Furthermore, a reduction in tensile strength and the displacement ratio D, as a parameter of the cycle loading, was reported because of the hardening of the binders. Figure 7.6 demonstrates the behavior of displacement ratio (*D*) with temperature for the different resin systems [20].

A vibration study of epoxy-based PC in various temperatures, 25°C, 50°C, 100°C, 150°C, 200°C, and 250°C, proved that the natural frequency and modulus of elasticity decreased by elevating the temperature [21].

Polymer binder plays a crucial role in the thermal behavior of PC. A study of the mechanical properties of the PC exposed to various temperatures up to 90°C presented that epoxy-based PC are more susceptible to thermal loads than the unsaturated polyester-based one because of the heat distortion temperature (HDT) of the binders [22].

Numerous works have been performed to increase the thermal stability of PC using different additive ingredients, such as micro fillers [23], fibers, and nanoparticles. In literature [14], authors investigated the influence of incorporating 2 wt% basalt fiber on the thermal stability of epoxy-based PC. The residual compressive and flexural strengths, as well as splitting tensile strengths of unreinforced PC, exhibited more reduction compared to basalt fiber-reinforced one at raised temperatures of 50°C, 100°C, 150°C, 200°C, and 250°C. Rising the temperature to 250°C decreased the value of residual compressive, flexural, and splitting tensile strengths of unreinforced PC by 54%, 50%, and 35%, respectively. On the other side, strength reduction

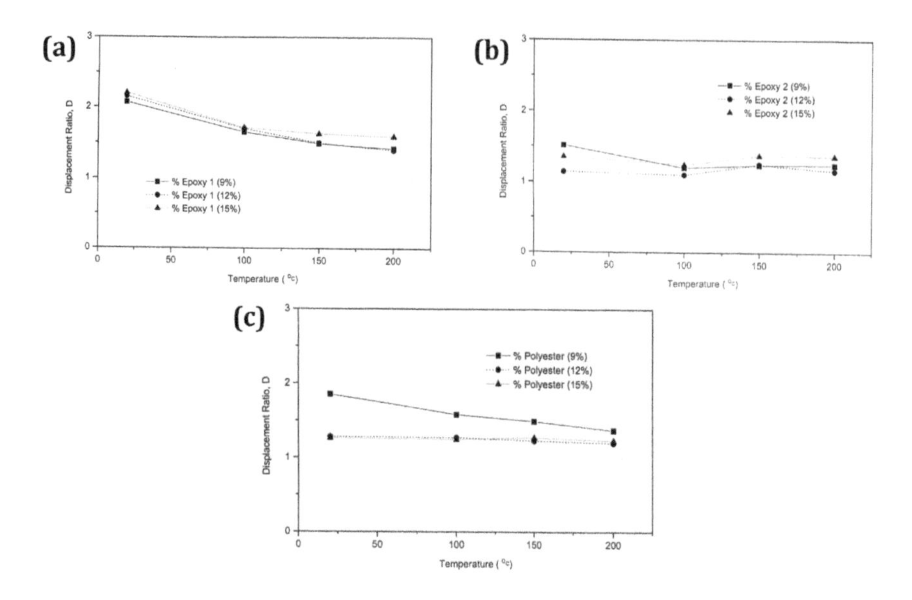

FIGURE 7.6 Variation of displacement ratio with temperature for the different resin systems: (a) grouting epoxy (epoxy 1), (b) solvent-free epoxy (epoxy 2), and (c) polyester [20]

in basalt fiber-reinforced PC did not observe by elevating the temperature up to 200°C. However, for the elevated temperature of 250°C, the compressive, flexural, and splitting tensile strengths were attained at about 77%, 88%, and 83% of their primary values, respectively. Therefore, chopped basalt fiber improved the thermal resistance of the PC. A little color change was reported for the samples subjected to 100°C, 150°C, and 200°C. The burning temperature of the basalt fiber-reinforced PC (BFRPC) specimen was reported at about 330°C. Both neat and reinforced PCs suffered weight loss after exposure to high temperatures. But the weight loss in neat PC was more. Another research demonstrated that incorporating basalt fiber into epoxy PC causes an increase in the compressive strength of the specimens after experiencing high temperatures of 100°C, 150°C, and 180°C [24].

For the assessment of the effect of flame retardant additives, two types of them, ammonium polyphosphate (APP) and an alumina trihydrate (ATH) were introduced to epoxy PC at a rate of 20 wt% of the polymer (13 wt% of the PC) to meliorate the thermo-mechanical properties of the PC [6]. The specimens were heated from 100°C to 250°C at a rate of 0.5°C/min and sustained for 3 hours, and then got cold at a rate of 0.5°C/min until the room temperature and tested after 24 hours. ATH flame retardant incorporation enhanced the rigidity of the PC before and after subjection to the target temperatures. The reduction of the residual mechanical strength of ATH-reinforced PC is lesser than the APP-reinforced PC and PC with optimal resin content. The APP-reinforced PC had the lowest modulus, and ATH-reinforced PC resulted in the most rigid PC for all subjected temperatures. The inequality of CTE of resin and aggregates and the formed cracks in PC due to chemical degradation of resin binder resulted in loss of matrix-aggregates bonding. This led to damage of the PC and reduced the elastic modulus. By raising the temperature from 23°C to 250°C porosity increased by 2%, which confirmed that the degradation of the PC was because of the chemical degradation of the binder and subjection of the cracks to the high and average temperatures. Increasing the temperature decreased the porosity of the APP-reinforced PC [6].

The addition of 2 wt% MMT nanoparticles improved the thermal resistance of epoxy-based PC. A comparative assessment of the residual compressive, flexural, and splitting tensile strengths of BFRPC and BFRPC incorporating 2 wt% nanoclay after experiencing the elevated temperatures are depicted in Figure 7.7 [14].

Heidarnezhad et al. [25] evaluated the effects of low temperatures on their PC performances. Reducing the temperature to −15°C, caused a decrease in the impact strength, energy absorption, and ductility of the PC, and also an increase in the compressive, splitting tensile, and flexural strengths as well as the elastic modulus of the PC.

7.4 FREEZE–THAW AND THERMAL CYCLE RESISTANCE

Some studies focused on the influence of different thermal cycles on PC performance. Exposure to the cyclic thermal loadings of 25–75°C during the 24-hour rapidly declined the bond strength of PC up to 45% after 80 thermal cycles. The temperature is gradually increased and decreased to avoid the thermal shock that is often associated with bridge deck overlays. Implementing fly ash micro filler slightly improved thermal cycling resistance by 7%. Authors attributed that to the superior workability of the PC and excellent placement quality in the casting process created by fly ash [23]. Evaluating the effect of thermal cycling ranging from approximately −16°C to 60°C on the interface shear stresses and longitudinal normal stresses of

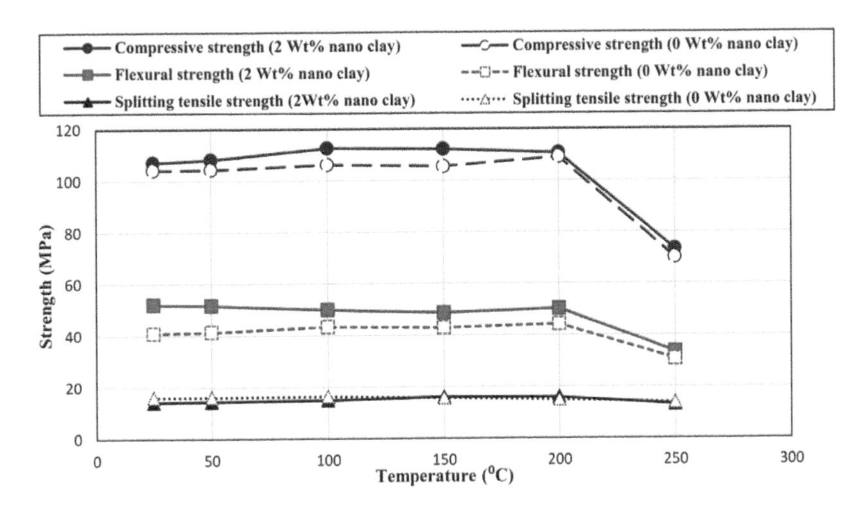

FIGURE 7.7 Influence of nanoclay reinforcement on the strength of basalt fiber-reinforced PC after experiencing high temperatures [14]

composite PC-PCC beams proved that interface shear stresses in simply supported composite beams were end bound [15].

Two types of study were conducted to determine the influence of a large range of temperatures, between –20°C and +100°C, on flexural strength of unsaturated polyester and epoxy PCs. In the first one, samples were heated (+5°C/min) and cooled (–2°C/min) to obtain the test temperatures and immediately tested. The second one was conducted as the first one, but samples were rapidly cooled by immersion in water at room temperature before the test. As can be seen in Figure 7.8, epoxy PC presented more susceptibility to temperature variation than polyester PC because of lower HDT of epoxy compared to polyester. After the HDT point, the flexural strength of epoxy PC decreased at a higher level. Flexural strength dropped by 14% and 75% after 50 and 100 thermal fatigue cycles between +20°C and 100°C, respectively. Besides, A little improvement of flexural strength was reported for epoxy PC at room temperature after sample tempering due to the post-curing effect. Freeze–thaw cycles (–10°C to +10°C) had no considerable influence on the flexural strength of both PCs. Also, one hundred freeze–thaw cycles formed just slight damage on both PCs [13].

Effects of three thermal cycles including 25°C to –30°C (cycle A), 25°C to 70°C (cycle B), and –30°C to 70°C (cycle C), applied for seven days to the PC samples, indicated that the interfacial shear strength influenced only by experiencing cycles A and B. The compressive and flexural strengths are not influenced by the freeze/thaw cycles (cycles A and C). Exposure to cycle B decreased the compressive and flexural strengths, and in fact, this cycle applied the most destructive influence. It was attributed to the low HDT point of the polymer binder [18]. Cycles A and B presented the maximum improvement and decline on tensile strength and K_{Ic}, respectively. In contrast, cycle C provides a mediocre influence. Rising the average temperature of

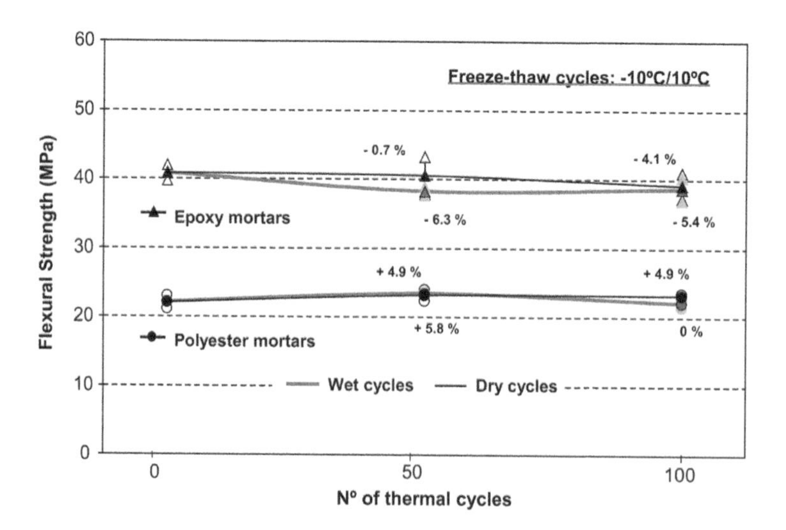

FIGURE 7.8 The behavior of residual flexural strength of epoxy and polyester PC after exposure to freeze–thaw cycles [13]

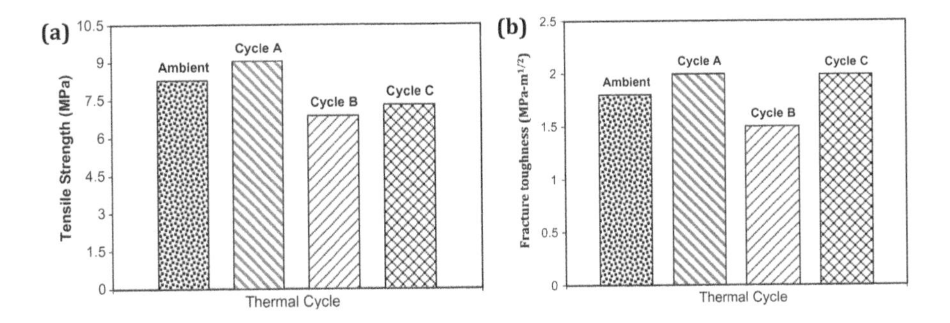

FIGURE 7.9 The behavior of (a) tensile strength and (b) fracture toughness of epoxy PC exposed to different cycles [26]

the cycles decreased both the K_{Ic} and tensile strength of the PC (Figure 7.9). Besides, cool-to-heat conditions might exhibit higher influences on tensile-type cracking of PC than heat-to-cool ones. Also, the heat-to-cool cycles increased the load-bearing capacity and persistence of the PC, while the cool-to-heat cycles extremely enhanced the risk of brittle tensile-type fracture [26].

Freeze–thaw resistance is one of the most significant properties which determines the resistance of concrete [27]. Freezing/thawing resistance was carried out on rubber-based PC reinforcing with polypropylene and steel fibers. For this purpose, static modulus of elasticity was used as a measured method to find the possible occurrence of damage after finishing 100, 200, 300, and 400 cycles. There were no considerable distinctions between the E0 and E400. However, the dynamic modulus of elasticity decreased by 25% after 400 cycles. So, the tested PC proved a high level of durability against freezing and thawing cycles [28]. The frost resistance of the PC composed of methyl methacrylate as a binder and quartz and as an aggregate was 100 freezing-thawing cycles. The specimens lost their masses by 2.0–4.5% after 100 cycles [29]. Also, it was reported that the time of freeze–thaw, lasting up to ten days, did not influence the mechanical properties of the elastic rubber PC [30].

Exposing to 50 frost-thaw cycles between −15°C and +20°C declined the compressive strength of epoxy polyurethane acryl PC by 15.58%. Naked eye analysis did not observe any changes in appearance (swelling, cracks, exfoliations, or material dislocations) of the PC [7]. Also, the thermal shock strength of the PC was carried out using the method of immersion, and the average water absorption was obtained at 0.0108%. From the visual examination, the PC did not show deformation, swelling, cracks, exfoliations, and material dislocations [7].

7.5 FIRE RESISTANCE AND FLAMMABILITY

A significant shortcoming of PC is the deficient manner under fire because of the inclusion of polymer binder. Since PC comprises high-weight content of filler and aggregates, its glass transition temperature and flammability are greatly increased [6].

Some approaches were applied for improving the fire resistance of PC. One of them is implementing fire resistance resins (e.g., phenolic resins), which provide lower mechanical strength. Another approach is the addition of fire retardant systems, which may be lead to toxicological impact in the environment due to great smoke generation. Also, incorporation of the high volume of fire retardant additives can decline the mechanical strength of PC. Using nanoparticles is another approach for improving the fire resistance of PC.

Fire reaction properties of PC was carried out using cone calorimeter test following ASTM E 1354 to study heat release rate (HRR), mass loss smoke, CO and CO_2 yield rates, time to ignition (IT), average specific extinction area (SEA), and effective heat of combustion. The experimental study demonstrated that nanodispersion of Al_2O_3 nanoparticles did not improve the fire resistance of epoxy PC. On the other side, the addition of 3.9 wt% Al_2O_3 dry powder improved fire reaction properties. Incorporating 3.9 wt% Al_2O_3 nanopowder reduced SEA, mass loss, peak, and average heat rate, respectively, by 16%, 15%, 9%, and 7%. Also, a significant improvement of about 22% in time to ignition was reported about plain PC. It was presented that the size reduction of Al_2O_3 particles from micro- to nano-improved flame retardancy of the PC [31,32].

Fire retardant filler is used to reduce the flammability of the PC due to its influences on the resistance to ignition and the smoke and toxic gas emission. They can also influence the heat capacity, thermal conductivity, and emissivity of PC. Although fire retardant filler is not entirely inert to fire, it will provide adequate time to react to the fire before any damage happens [33].

A flammability test was carried out according to ASTM D 635-97 to determine the burning characteristics and revealed that the PCs are not flammable [34]. However, the unsaturated polyester PC burned with smoke and flammable drips. Marble waste and basalt improve the flame retardancy of PC [34].

7.6 HOT WATER RESISTANCE

The hot water resistance of PC should be considered to investigate the heat resistance and the changes in strength before and after exposure to hot water. For this purpose, hot water (90°C) immersion test was conducted for four weeks and proved the remarkable negative influence of hot water on compressive and flexural strengths [35]. The decline rate of flexural strength is much higher compared to compressive strength. During the test, pores and cracking were generated by the decomposition and degradation of the matrix. On the other hand, hot water decreased the bulk density and increased the porosity of the PC due to polymer decomposition [35].

From the experimental study in literature [36], increasing the immersion time in boiling water raised the erosion depth in polyester PC. Although the appearance and weight of the specimens did not change, the compressive and tensile strengths decreased.

REFERENCES

[1] D.N. O'Connor, M. Saiidi, Compatibility of polyester-styrene polymer concrete overlays with Portland cement concrete bridge decks, *ACI Mater. J.* 90 (1993) 59–68.
[2] K.S. Rebeiz, Time-temperature properties of polymer concrete using recycled PET, *Cem. Concr. Compos.* 17 (1995) 119–124. https://doi.org/10.1016/0958-9465(94)00004-I.

[3] H. Haddad, M. Al Kobaisi, Optimization of the polymer concrete used for manufacturing bases for precision tool machines, *Compos. Part B Eng.* 43 (2012) 3061–3068. https://doi.org/10.1016/j.compositesb.2012.05.003.
[4] M.C.S. Ribeiro, J.M.L. Reis, A.J.M. Ferreira, A.T. Marques, Thermal expansion of epoxy and polyester polymer mortars—plain mortars and fibre-reinforced mortars, *Polym. Test.* 22 (2003) 849–857. https://doi.org/10.1016/S0142-9418(03)00021-7.
[5] H. Haddad, M. Al, Influence of moisture content on the thermal and mechanical properties and curing behavior of polymeric matrix and polymer concrete composite, *Mater. Des.* 49 (2013) 850–856. https://doi.org/10.1016/j.matdes.2013.01.075.
[6] O. Elalaoui, E. Ghorbel, M. Benouezdou, The effect of the fire flame retardant type on the durability of polymer concrete exposed to elevated temperature, in: J.B. de Aguiar, S. Jalali, A. Camoes, R.M. Ferreira (Eds.), *Int. Congr. Polym. Concr. (ICPIC)*, Funchal-Maderia, Portugal, 2010: pp. 127–132.
[7] L. Agavriloaie, S. Oprea, M. Barbuta, F. Luca, Characterisation of polymer concrete with epoxy polyurethane acryl matrix, *Constr. Build. Mater.* 37 (2012) 190–196. https://doi.org/10.1016/j.conbuildmat.2012.07.037.
[8] O. Elalaoui, E. Ghorbel, V. Mignot, M. Ben Ouezdou, Mechanical and physical properties of epoxy polymer concrete after exposure to temperatures up to 250°C, *Constr. Build. Mater.* 27 (2012) 415–424. https://doi.org/10.1016/j.conbuildmat.2011.07.027.
[9] G.B. Ramesh, U.G. Student, Review on performance of polymer concrete with resins and its applications, *Int. J. Pure Appl. Math.* 119 (2018) 175–184.
[10] A. Kumar, G. Singh, N. Bala, Evaluation of flexural strength of epoxy polymer concrete with red mud and fly ash, *Int. J. Curr. Eng. Technol.* 13 (2013) 1799–1803.
[11] K. Aniskevich, J. Hristova, Creep of polyester resin filled with minerals, *J. Appl. Polym. Sci.* 77 (2000) 45–52.
[12] W.N. Findley, *Creep and Relaxation of Nonlinear Viscoelastic Materials*, 1st ed., Dover publication, New York, 1976.
[13] A.J.M. Ferreira, A.T. Marques, M.C.S. Ribeiro, P.R. N, Flexural performance of polyester and epoxy polymer mortars under severe thermal conditions, *Cem. Concr. Compos.* 26 (2004) 803–809.
[14] M.H. Niaki, A. Fereidoon, M.G. Ahangari, Experimental study on the mechanical and thermal properties of basalt fiber and nanoclay reinforced polymer concrete, *Compos. Struct.* 191 (2018) 231–238. https://doi.org/10.1016/j.compstruct.2018.02.063.
[15] B.D.L. Wheat, D.W. Fowler, A.I. Ai-negheimish, Thermal and fatigue behavior of polymer concrete overlaid beams, *J. Mater. Civ. Eng.* 5 (1993) 460–477.
[16] M. Hassani, N. Abdolhosein, F. Morteza, G. Ahangari, Mechanical properties of epoxy/basalt polymer concrete: Experimental and analytical study, *Struct. Concr.* 19 (2018) 366–373. https://doi.org/10.1002/suco.201700003.
[17] W. Ferdous, A. Manalo, H.S. Wong, R. Abousnina, O.S. Alajarmeh, Y. Zhuge, P. Schubel, Optimal design for epoxy polymer concrete based on mechanical properties and durability aspects, *Constr. Build. Mater.* 232 (2020) 117229. https://doi.org/10.1016/j.conbuildmat.2019.117229.
[18] M.M. Shokrieh, M. Heidari-Rarani, M. Shakouri, E. Kashizadeh, Effects of thermal cycles on mechanical properties of an optimized polymer concrete, *Constr. Build. Mater.* 25 (2011) 3540–3549. https://doi.org/10.1016/j.conbuildmat.2011.03.047.
[19] A. Blaga, J. Beaudoin, Polyme concrete, *Can. Build. Dig.* CBD-242. (1985) 1–4.
[20] M.M. El-Hawary, H. Abdel-Fattah, Temperature effect on the mechanical behavior of resin concrete, *Constr. Build. Mater.* 14 (2000) 317–323. https://doi.org/10.1038/laban0208-61a.
[21] B. Imane, L. Boudjemaa, B. Ibtissam, Mechanical characterization of resin concrete subjected to high temperatures by vibration analysis, *Mater. Sci. Forum.* 962 (2019) 236–241. https://doi.org/10.4028/www.scientific.net/MSF.962.236.

[22] J.M.L. dos Reis, Effect of temperature on the mechanical properties of polymer mortars, *Mater. Res.* 15 (2012) 645–649. https://doi.org/10.1590/S1516-14392012005000091.
[23] K. Rebeiz, S. Serhal, A. Craft, Properties of polymer concrete using fly ash, *J. Mater. Civ. Eng.* 16 (2004) 15–19. https://doi.org/10.1061/(ASCE)0899-1561(2004)16:1(15).
[24] S.E. Mohammadyan-Yasouj, H.A. Ahangar, N.A. Oskoei, H. Shokravi, S.S.R. Koloor, M. Petrů, Experimental study on the effect of basalt fiber and sodium alginate in polymer concrete exposed to elevated temperature, *Processes.* 9 (2021). https://doi.org/10.3390/pr9030510.
[25] F. Heidarnezhad, K. Jafari, T. Ozbakkaloglu, Effect of polymer content and temperature on mechanical properties of lightweight polymer concrete, *Constr. Build. Mater.* 260 (2020) 119853. https://doi.org/10.1016/j.conbuildmat.2020.119853.
[26] M. Heidari-Rarani, M.R.M. Aliha, M.M. Shokrieh, M.R. Ayatollahi, Mechanical durability of an optimized polymer concrete under various thermal cyclic loadings – An experimental study, *Constr. Build. Mater.* 64 (2014) 308–315. https://doi.org/10.1016/j.conbuildmat.2014.04.031.
[27] R. Allahvirdizadeh, R. Rashetnia, A. Dousti, M. Shekarchi, Application of the polymer concrete in repair of concrete structures : A literature review, in: *4th Int. Conf. Concr. Repair,* Dresden, Germany, 2011: pp. 1–10. https://doi.org/10.13140/2.1.4893.7925.
[28] J. Sustersic, A. Zajc, I. Leskovar, V. Dobnikar, PC and LMC with granulated rubber made from waste car tires, in: J.B. de Aguiar, S. Jalali, A. Camoes, R.M. Ferreira (Eds.), *13th Int. Congr. Polym. Concr. (ICPIC),* Funchal-Maderia, Portugal, 2010: pp. 289–297.
[29] L. Trykoz, S. Kamchatnaya, O. Pustovoitova, A. Atynian, O. Saiapin, Effective waterproofing of railway culvert pipes, *Balt. J. Road Bridg. Eng.* 14 (2019) 473–483. https://doi.org/https://doi.org/10.7250/bjrbe.2019-14.453.
[30] K. Chung, Y. Hong, Weathering properties of elastic rubber concrete comprising waste tire solution, *Polym. Eng. Sci.* 49 (2009) 794–798. https://doi.org/10.1002/pen.
[31] M.C.S. Riberio, C.M.C. Pereira, M. Martins, A. Marques, A.J.M. Ferreira, Effects of micro and nano-sized Al_2O_3 particles on mechanical behaviour and fire reactuion properties of wpoxy polymer concrete, in: *Int. Congr. Polym. Concr.*, Univ. of Minho, Dep. of Civil Engineering, Funchal, 2010: pp. 257–265.
[32] A.J.M.F.M.C.S. Riberio, C.M.C. Pereira, S.P.B. Sousa, P.R.O. Novoa, Fire reaction and mechanical performance analyses of polymer concrete materials modified with micro and nano alumina particles, *Restor. Build. Monum.* 19 (2013) 195–202.
[33] A. Beutel, *Optimal Mix Design for Epoxy Resin Polymer Concrete*, University of Southern Queensland, Darling Heights, 2015.
[34] M.E. Tawfik, S.B. Eskander, Polymer concrete from marble wastes and recycled poly(ethylene terephthalate), *J. Elastomers Plast.* 38 (2006) 65–79. https://doi.org/10.1177/0095244306055569.
[35] E. Hwang, J. Kim, S. Park, Physical properties of polyester polymer concrete composite using RCSS fine aggregate, *Adv. Mater. Res.* 687 (2013) 219–228. https://doi.org/10.4028/www.scientific.net/AMR.687.219.
[36] Y. Ohama, Hot water resistance of polyester resin concrete, in: *20th Japan Congr. Mater. Resist.*, Tokyo, Japan, 1977: pp. 176–178.

8 Chemical Resistance of Polymer Concrete

Abstract

This chapter studies the chemical resistance of polymer concrete (PC), stability against corrosive environments, and chemical attacks to maintain its properties as well as appearance. One valuable property of PC is its good chemical resistance. Usually, PCs are more resistant in aggressive environments than ordinary Portland cement concretes, because of incorporating polymer binder, but it is not always true. In this chapter, in addition to examining the chemical stability of PC against a variety of chemical solutions, acids and bases, corrosive environments, and chemical attacks, the effect of exposure to these conditions on the mechanical properties such as compressive, flexural, and tensile strengths of PC are discussed. Also based on previous works some factors for improvement of their chemical resistance are presented.

A comparative assessment of the long-term durability of epoxy polymer concrete (PC) and ordinary Portland cement (OPC) concrete in different aggressive mediums presented that the PC had better resistance. PC specimens with high epoxy resin content showed excellent chemical resistance [1]. Thus, it is necessary to select the suitable materials (polymer binder and aggregates) and weight contents for PC related to chemically aggressive environments [2].

Aggregate and polymer selection influence the chemical stability of PC. Polymers are relatively chemically inactive materials. The chemical resistance of PCs has been studied in the literature that frequently accounts for corrosion in industrial environments, seawater, and so on. PCs exhibit good chemical stability to corrosive media, especially when compared to OPC concretes [3]. Most PCs are resistant to alkalis, acids, and a wide range of other aggressive media such as ammonia, petroleum products, salt environments, and some solvents. According to ASTM C267, four important parameters which should be evaluated in chemical resistance study are weight change percentage, compressive strength change percentage, the appearance of the specimen, and appearance of the test medium [4].

Golestaneh et al. [5] evaluated the chemical stability of epoxy PC for 7, 14, 28, and 56 days of exposure in 15, 30, and 60 wt% concentrations of different aggressive media. Increasing the concentration up to 60% caused surface whitening and slag formation in sodium hydroxide (NaOH) solution, color changes in citric acid and hydrochloric acid solutions, swell, and destruction in sulfuric acid (H_2SO_4) and acetic acid solutions. The acetic acid solution caused a terrible reduction in compressive strength and significant weight loss in specimens. Furthermore, all the specimens except those in acetic acid solution slightly gained some weight after a long duration of immersion in aggressive media. As a result, the PC samples had excellent chemical stability in all chemicals except acetic acid solution.

DOI: 10.1201/9781003326311-8

A chemical resistance test was carried out on vinyl ester concretes incorporating waste mineral dust in the literature [6]. After 31 days of exposure in 1 M sulfuric acid or 4% sodium base, the mass and compressive strength variations were measured. Increasing the polymer content led to reduced PC strength after the chemical attack. On the other side, applying more dust in the micro filler out came a more minor reduction. It was concluded that the waste mineral dust did not considerably decrease the chemical stability. The presented PC demonstrated more sensitivity to sodium base compared to H_2SO_4. The optimal composition for counteracting the mass loss was the PC with a higher polymer content due to the good coverage of the micro filler particles. On the other side, implementing less polymer in the PC led to a smaller decline in compressive strength after a NaOH exposure.

From the statistical analysis [3], the type of resin, micro filler content, type of chemical solution, and interaction of mentioned elements considerably influenced the flexural strength of PC subjected to chemical attack. The acid environments (acetic acid, citric acid, formic acid, lactic acid, sulfuric acid, cola soft drink, and distilled water) did not change the surface and weight of PC. However, the flexural strength of the PC decreased. The strength reduction was more evident in the combinations with lower filler amounts. The lowest inclusion of fly ash (8 wt%) in both isophthalic and orthoptic polyester PC led to the lowest resistance to chemical agents. By increasing the fly ash content, the number of voids in the PC decreased, and therefore, better packing was achieved, which reduced the penetration of aggressive agents. The aggressive solutions penetrated through the pore network of high porosity PC (8 wt% fly ash) reached the binder–aggregate interface and weakened the polymer/aggregate bonding, as illustrated in Figure 8.1 [3].

Polyester PC is sensitive to deterioration in aggressive media, especially alkaline solutions. Fifteen days immersion in alkali solution caused a substantial drop in strength. Also, immersion in NaOH solution reduced the weight by 0.5%.

In literature [7], influence of high-concentration alkaline solution (50% NaOH) and sulfuric acid (16% H_2SO_4) on unsaturated polyester PC was examined. In an alkaline solution, the weight loss is greater than in an acid test. It was observed that acid and alkaline solutions attacked the interfacial regions as well as the polymer binder, and therefore accelerated the deterioration process. The alkaline solution caused the most important physical surface changes and reduced compressive and flexural strengths. It can be concluded that exposure to alkali environments constitutes more severe environmental conditions when compared to acid environments.

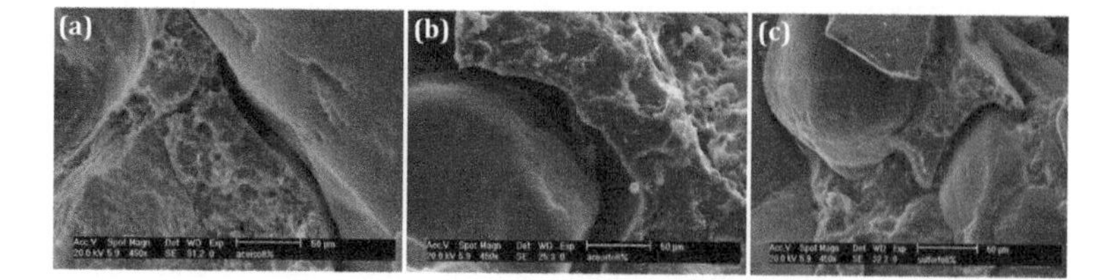

FIGURE 8.1 Polymer concretes incorporating 8% fly ash: (a) isophthalic PC in acetic acid, (b) orthophtalic PC in acetic acid, and (c) orthophtalic PC in sulfuric acid [3]

Another study considered the behavior of polyester PC to keep physical (mass change) and mechanical properties (flexural and compressive strengths) subjected to an alkaline attack by potassium hydroxide of a concentration of 5% [8]. As a result, incorporating more perlite powder causes decrease in the durability of the PC. It was reported that the composition with the lower apparent density, and thus higher porosity, led to the deeper penetration of the aggressive agent into the PC.

In Ref. [9], the chemical stability of epoxy and polyester PC is determined by the variation of mass and flexural strength after immersing in water solutions of sulfuric acid and sodium chloride (both are ordinary aggressive media), for different periods from 1 to 84 days. From the experiments, the immersion of epoxy PC in both solutions influenced the flexural strength slightly, which indicated the considerable chemical stability of this kind of PC. Both concretes exhibited minimum flexural strength reduction in sodium chloride solution. On the other side, increasing the immersion period (days) generally increased the mass change and flexural strength decrement of polyester PC. Also, increasing the water uptake causes decrease in the flexural strength. The different chemical resistance of the two kinds of PCs, attributed to the resin binder permeability characteristics [9]. It was found that elastic rubber concrete was not affected by sodium chloride and H_2SO_4 for up to 240 days [10].

A comprehensive study according to ASTM D543-06 on the effects of tap water, groundwater, Mediterranean seawater, 20% Na_2CO_3, 10% NaOH, and 10% H_2SO_4 revealed that all specimens retained their integrity and appearance and no swelling, and almost no weight loss was detected. Also, no identified decrease in the flexural strength, excluding a reduction in the PC subjected to 10% NaOH and 10% H_2SO_4 was reported. Immersing in 10% H_2SO_4 resulted in color change and the formation of small pores in the PC specimen. The studied PC had slight water absorption during four weeks of immersion and relatively remarkable chemical stability [11].

A comparative investigation of chemical resistance between OPC concrete and polyester PC proved that the immersion of the PC in 10% hydrochloric acid and 10% H_2SO_4 for 28 days did not result in weight loss. However, OPC concrete presented a weight loss of about 50% [12].

Mebarkia and Vipulanandan [13] evaluate compressive and splitting tensile strengths of polyester PC after 30 days of exposure to sulfuric acid, sodium hydroxide, and sodium chloride solutions with different pH values. All solutions reduced the strength of the PC, and by raising the pH level of the chemical solutions, the strength of PC declined. The type, concentration, and pH of the corrosive solutions are the important parameters that influence the strength of the PC. However, the PC specimens did not exhibit any visible deterioration after a one-month immersion.

Studying the effects of 11 typical reagents (hydrochloric acid, sulfuric acid, acetic acid, sodium hydroxide, sodium sulfate, sodium chloride, kerosene, rapeseed oil, toluene, acetone, and tap water) on the compressive strength of PMMA PC proved the remarkable resistance to tapping water, alkali, salts, kerosene, and rapeseed oil. However, the PC is significantly degraded by acetone (Figure 8.2a) and toluene (Figure 8.2b) (due to the inferior resistance of applied resin) and slightly attacked by acids (because of reactions with ground calcium carbonate incorporated as a micro filler) [14].

The behavior of vinyl and polyester PC as a cover of the bridge deck, when subjected to water and chemicals like motor oil and antifreeze solutions, was considered

FIGURE 8.2 Polymethyl methacrylate concrete immersed in (a) toluene for 28 days and (b) acetone for 28 days [14]

in the literature [15]. It was observed that water absorbent caused weakness of the interface between the polymer binder and the aggregate, which led to degradation of the PC.

Another research evaluated the effect of lactic acid, sulfuric acid, cola soft drink, and distilled water on the performance of epoxy PC. Submitting to sulfuric acid decreased the flexural strength by 10.41%, and a reduction of 6.87% and 5.54% were reported after exposure to seawater and distilled water, respectively. The compressive strength of PC decreased by 7.0%, 2.1%, and 1.7% when submitted to seawater, sulfuric acid, and distilled water, respectively. Also, the rate of degradation of the sample in lactic acid was such that the tests were not feasible. So, the flexural strength reduction was more in comparison with the compressive strength of the PC. Sulfuric and lactic acids cause change in the appearance of the PC, but no significant weight loss was reported. The loss of strength can be addressed not to the degradation of the polymer binder used in the PC, but to an enhancement of porosity in specimens with extended capillary diffusion of aggressive solutions, which diminishes the polymer/aggregate bonding strength [16].

Chikhradze et al. [17] introduced corrosion-resistant PC reinforced with basalt and steel fibers and compared shock resistance of the PC and cement concrete after one year of exposure to two different aggressive media. Shock resistance of the PC was obtained 31.2 kgf.cm/cm^3, about 624 times higher than the cement concrete, in a 5% Na_2SO_4 medium. Also, it was obtained 28.3 kgf.cm/cm^3, about 2,830 times higher than the cement concrete, in a 1.5% H_2SO_4 medium. An assessment of the influence of basalt and steel fiber reinforcement on corrosion resistance of polyester PC revealed that the steel fiber has more susceptibility to corrosive media [17].

Resistance to H_2SO_4 corrosion of vinyl ester-based PC was studied in Ref. [18]. After 50 days of acid immersion, the compressive strength did not change. However, the flexural strength of the PC decreased slightly. Scanning electron microscopy analysis demonstrated that the inner structure of the samples was unchanged after the exposure to H_2SO_4. However, the PC surfaces deteriorated because of acid corrosion. In applications where high chemical resistance is required, metallurgical aggregates can be a good alternative to natural aggregates [19].

According to the research results, the effective parameters for increasing the chemical resistance of PC can be listed as follows:

1. Incorporating more polymer weight content: The use of more resin in the composition causes better coverage of aggregates and fillers and reduces the penetration of chemicals into the PC. Also, the chemical resistance of polymers is often higher than that of aggregates and fillers.
2. Reduce PC water absorption: Water absorption weakens the interface between the polymer and the aggregate, leading to PC degradation.
3. Increasing the apparent density of PC: Higher density reduces the permeability of chemicals to PC and increases chemical stability.
4. Decreasing the porosity of PC: Reducing the porosity of the PC structure reduces the penetration depth of chemical agents into the concrete and increases the chemical resistance.
5. Using micro fillers: The inclusion of micro fillers such as fly ash significantly decreases the microvoids in PC and leads to better packing. Thus, the aggressive agents' penetration decreases.
6. Implementing chemical resistance fibers: Some fibers, such as basalt fibers, which have significant chemical resistance, can improve the chemical stability of PC.
7. Using metallurgical aggregates: Metallurgical aggregates are more resistant to some chemical environments than natural aggregates and can be a good alternative.

REFERENCES

[1] P. Ghassemi, V. Toufigh, Durability of epoxy polymer and ordinary cement concrete in aggressive environments, *Constr. Build. Mater.* 234 (2020) 117887. https://doi.org/10.1016/j.conbuildmat.2019.117887.

[2] R. Allahvirdizadeh, R. Rashetnia, A. Dousti, M. Shekarchi, Application of the polymer concrete in repair of concrete structures : A literature review, in: *4th Int. Conf. Concr. Repair*, Dresden, Germany, 2011: pp. 1–10. https://doi.org/10.13140/2.1.4893.7925.

[3] J.P. Gorninski, D.C. Dal Molin, C.S. Kazmierczak, Strength degradation of polymer concrete in acidic environments, *Cem. Concr. Compos.* 29 (2007) 637–645. https://doi.org/10.1016/j.cemconcomp.2007.04.001.

[4] C. Astm– 01(2012) standard test methods for chemical resistance of mortars, grouts, and monolithic surfacings and polymer concretes, *ASTM Int.* C267-01 (2012) 1–6. https://doi.org/10.1520/C0267-01R12.

[5] M. Golestaneh, G. Najafpour, G. Amini, M. Beygi, Evaluation of chemical resistance of polymer concrete in corrosive environments, *Iran. J. Energy Environ.* 4 (2013) 304–310. https://doi.org/10.5829/idosi.ijee.2013.04.03.19.

[6] J.J. Sokołowska, P.P. Woyciechowski, Chemical resistance of vinyl-ester concrete with waste mineral dust Remaining after preparation of aggregate for asphalt mixture, in: M.M.R. Taha (Ed.), *Int. Congr. Polym. Concr. (ICPIC)*, Springer, Cham, Washington, DC, 2018: pp. 491–497.

[7] M. Robles, S. Galan, Durability of polyester polymer concrete under the influence of chemical solutions, in: J.B. de Aguiar, S. Jalali, A. Camoes, R.M. Ferreira (Eds.), *Int. Congr. Polym. Concr. (ICPIC)*, Funchal-Maderia, Portugal, 2010.
[8] S.J. Julia, W.P. Paweł, Ł. Paweł, K. Kamila, Effect of perlite waste powder on chemical resistance of polymer concrete composites, *Adv. Mater. Res.* 1129 (2015) 516–522. https://doi.org/10.4028/www.scientific.net/AMR.1129.516.
[9] M.C.S. Ribeiro, C.M.L. Tavares, A.J.M. Ferreira, Chemical resistance of epoxy and polyester polymer concrete to acids and salts, *J. Polym. Eng.* 22 (2011) 27–44. https://doi.org/https://doi.org/10.1515/POLYENG.2002.22.1.27.
[10] K. Chung, Y. Hong, Weathering properties of elastic rubber concrete comprising waste tire solution, *Polym. Eng. Sci.* 49 (2009) 794–798. https://doi.org/10.1002/pen.
[11] K. Shi-cong, P. Chi-sun, A novel polymer concrete made with recycled glass aggregates, fly ash and metakaolin, *Constr. Build. Mater.* 41 (2013) 146–151. https://doi.org/10.1016/j.conbuildmat.2012.11.083.
[12] T. Yamamoto, The production performance and potential of polymers in concrete, in: *Int. Congr. Polym. Concr.*, Brighton, England, 1987: pp. 395–398.
[13] S. Mebarkia, C. Vipulanandan, Mechanical properties and water diffusion in polyester polymer concrete, *J. Eng. Mech.* 121 (1995) 1359–1365. https://doi.org/https://doi.org/10.1061/(ASCE)0733-9399(1995)121:12(1359).
[14] Y. Ohama, T. Kobayashi, K. Takeuchf, K. Nawata, Chemical resistance of polymethyl methacrylate concrete, *Int. J. Cem. Compos. Light. Concr.* 8 (1986) 87–91. https://doi.org/https://doi.org/10.1016/0262-5075(86)90003-5.
[15] B. Sarde, Y.D. Patil, Recent research status on polymer composite used in concrete-an overview. *Mater. Today: Proc.* 18 (2019) 3780–3790. https://doi.org/10.1016/j.matpr.2019.07.316.
[16] J.M.L. Reis, A comparative assessment of polymer concrete strength after degradation cycles, in: H. da C. Mattos, M. Alves (Eds.), *Second Int. Symp. Solid Mech.*, Brazilian Society of Mechanical Sciences and Engineering, Rio de Janeiro, Brazil, 2009: pp. 437–444.
[17] N. Chikhradze, F. Marquis, G. Abashidze, D. Tsverava, Production of corrosion-resistant polymer concrete reinforced with various fibers, in: *IOP Conf. Ser. Earth Environ. Sci. 362, IOP Conf.*, 2019. https://doi.org/10.1088/1755-1315/362/1/012118.
[18] M.S. Yinong Shen, B. Liu, J. Lv, Mechanical properties and resistance to acid corrosion of polymer concrete incorporating ceramsite, fly ash and glass fibers, *Materials (Basel).* 12 (2019) 2441. https://doi.org/10.3390/ma12152441.
[19] A. Seco, A.M. Echeverría, S. Marcelino, B. García, S. Espuelas, Durability of polyester polymer concretes based on metallurgical wastes for the manufacture of construction and building products, *Constr. Build. Mater.* 240 (2020) 117907. https://doi.org/10.1016/j.conbuildmat.2019.117907.

9 Some Other Properties of Polymer Concrete

Abstract

This chapter evaluates ageing effect, weathering resistance, ultraviolet (UV) and gamma radiation, and dielectric and insulation properties of polymer concrete (PC), which have received less attention in previous research. First, the effect of ageing on the strength of PC is discussed. Due to the exposure of PC structures in outdoor conditions in some applications, it is necessary to study its resistance to various weather conditions such as day and night and spring and summer weather conditions. Furthermore, due to the exposure to sunlight, the study of the destructive effect of UV and gamma radiation on the properties of PC is also important. Because of their insulation properties, PCs have many applications in the electrical and communication industries. The electrical resistance and the parameters affecting it are also discussed in this chapter.

9.1 AGEING EFFECT

Like cement concretes, the properties of polymer concretes (PCs) change over time. In the study of curing effects, it was noted that the changes in properties in the early hours and days are significant. The compressive strength quickly increases up to the curing age of 24 hours and then slowly increases up to the curing age of 72 hours [1]. Compressive strength variations of polyester, epoxy, and vinyl ester PCs at 3, 7, 14, 21, and 28 days of curing indicated that they keep on achieving strength with time, as demonstrated in Figure 9.1.

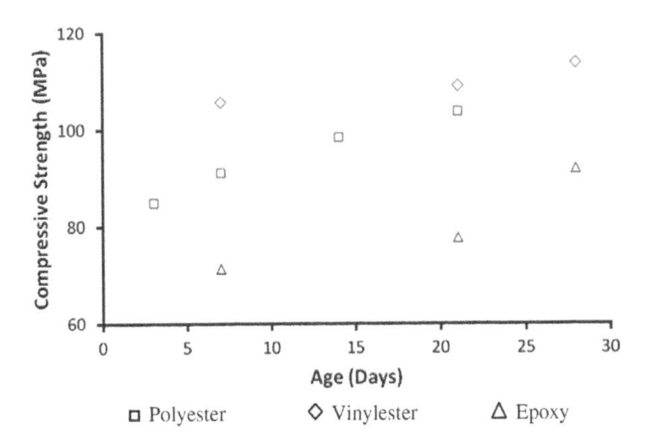

FIGURE 9.1 Compressive strength variation with age [2]

DOI: 10.1201/9781003326311-9

Variation of strength with time for the epoxy PC is maximum, while it was minimum for vinyl ester PC. It is observed that all the PCs achieved about 80% of the 28-day strength, only after seven days of curing [2]. PCs may continue to increase their strength for years until they reach their final strength. Long-term strength study of low-viscosity vinyl ester PC demonstrated a significant increment in the compressive strength after seven years. It was estimated that a period of nine years would be sufficient to achieve the final strength of PC [3]. This research confirms the high durability of PCs. Also, 7-year-old vinyl ester PC incorporating different fly ash content exhibited higher compressive strength than the 14-day-old one [4].

Ma et al. [5] experimentally determined the influence of ageing on residual flexural strength of epoxy PC. An accelerating ageing test was conducted in a weathering test chamber that provided ultraviolet (UV)-light irradiation, periodic variation of temperature (simulation of day and night), and weather humidity program according to the outdoor conditions in South China. It was observed that by increasing the ageing time equivalent to four years, the flexural strength of the PC declined about 8.4%.

9.2 WEATHERING RESISTANCE

Weathering conditions may be affected on PC properties. Literature almost considered only the influences of freeze–thaw and thermal cycles on PC performance, as previously explained. During the exposure in the outdoor environment, some parameters may affect the PC performance, such as light, UV, O_2, O_3, humidity, rain, and snow.

Figure 9.2 depicts the compressive stress–strain diagram of the epoxy PC after long-term exposure to air with 20°C and 30% humidity. Although the duration of exposure has little effect on the initial slope of the stress–strain curve, its effect on the strength of concrete is significant. The results of this study show that the strength of concrete increases to 33% after four months and does not change significantly after that. Also, one year of exposure to air increased the weight of concrete by

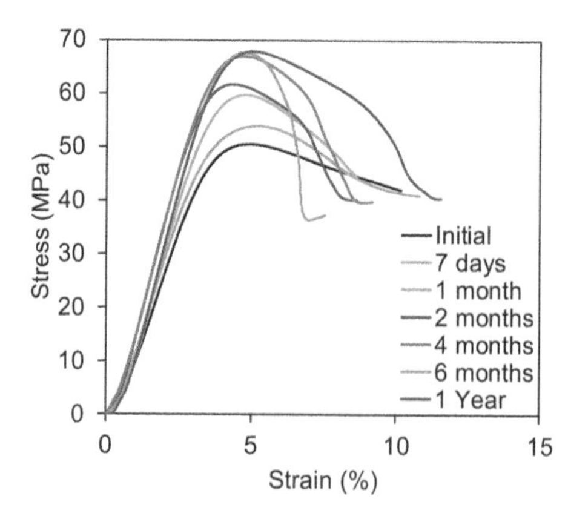

FIGURE 9.2 Compressive stress–strain diagram of the epoxy PC after up to 365 days of exposure in air [6]

0.13%. This slight weight gain can be attributed to the small absorption of water molecules in the air [6].

The destructive effect of spring and summer weather conditions on the flexural strength of PC is two to five times higher than in the autumn and winter seasons. Further degradation in hot seasons can be attributed to heat from more intense sunlight as well as UV radiation [7].

Long-term outdoor exposure is expected to reduce the flexural strength of epoxy and polyester concrete by 50% and 35%, respectively. However, the reduced flexural strength of PC after long-term exposure is still greater than the flexural strength of conventional cement concrete that is not exposed to the outdoor condition. Exposure to air with moisture cycles reduces the flexural strength of PC in a non-linear manner and accelerates the deterioration process up to 3.97 times [7].

Weathering properties (outdoor condition for up to 720 days) of elastic rubber concrete (ERC) were studied in Ref. [8], and no changes in the mechanical properties were observed.

9.3 ULTRAVIOLET AND GAMMA RADIATION

Polymer composites may degrade by UV radiation during their lifetime. Degradation occurs due to the break of polymer bonds and the oxidation process and is usually accompanied by discoloration of the PC [7]. It is important to evaluate the durability of PC structures exposed to UV radiation. However, not much research has been concentrated on the influence of UV radiation on PCs so far.

To evaluate the influence of UV radiation on glass fiber-reinforced polyester PC, the samples were subjected to UV radiation of wavelength 300 ± 400 nm for zero to four weeks. The authors selected 300 nm as the minimum radiation wavelength because a shorter wavelength is mostly absorbed by the ozone layer. As presented in Figure 9.3, the PC is sensitive to UV radiation. Accelerated tests revealed a strength reduction of about 25% after eight weeks [9]. It was attributed to the deterioration of the polymer by photo-oxidation of groupings on the polyester chains, which led to the brittle manner of the PC [9].

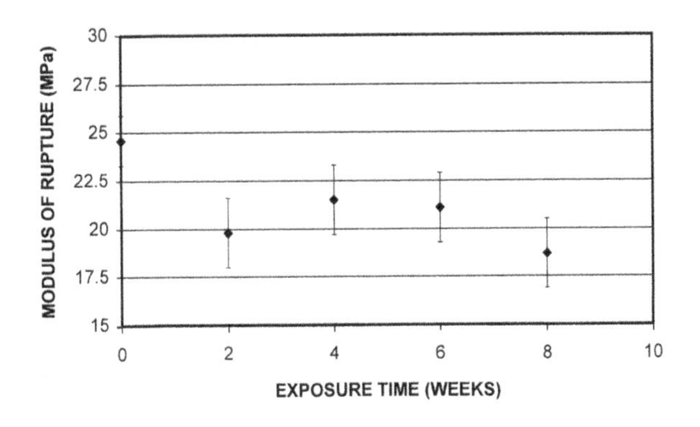

FIGURE 9.3 Degradation of modulus of rupture after exposure to UV radiation [9]

The resistance against UV radiation is one of the limitations of epoxy, which can be resolved by introducing fly ash filler [10]. The cycloaliphatic epoxy resin could be implemented for outdoor insulation purposes due to the significantly better UV resistance compared to an aromatic resin [11].

Gonzalo Martı́nez-Barrera et al. [12] investigated the effects of gamma radiation (0–100 kGy radiation dose) on nylon fiber-reinforced PC based on unsaturated pre-accelerated polyester resin. From the experiments, all irradiated PC samples have higher compressive strength and strain at yield point values than those non-irradiated. However, in the case of compression elastic modulus, a different trend is observed for PC with 0.3 or 0.4 vol% nylon fiber. Another research demonstrated that UV radiation did not decrease the flexural strength of the epoxy-based PC [5].

Figure 9.4 represents the effect of gamma radiation dose on the compressive strength of the polyester-based PC containing waste Tetra Pack particles. The control PC (without additive) has the maximum compressive strength. Exposing to the gamma irradiation up to 100 kGy increases the compressive strength of all specimens, and the maximum compressive strength values were obtained by 100 kGy gamma radiation dose. By increasing the irradiation dosage, the compressive strength values decreased [13].

Also, the flexural strength of the concrete in the face of gamma irradiation, has shown a behavior quite similar to the compressive strength. As can be observed in Figure 9.5, exposure to the 100 kGy increases the flexural strength of all PC samples [13].

9.4 DIELECTRIC AND INSULATION PROPERTIES

The history of research on the use of PCs in the electrical industry dates back to the 1970s by Westinghouse R&D Center in Pittsburgh, PA [14]. Due to high energy

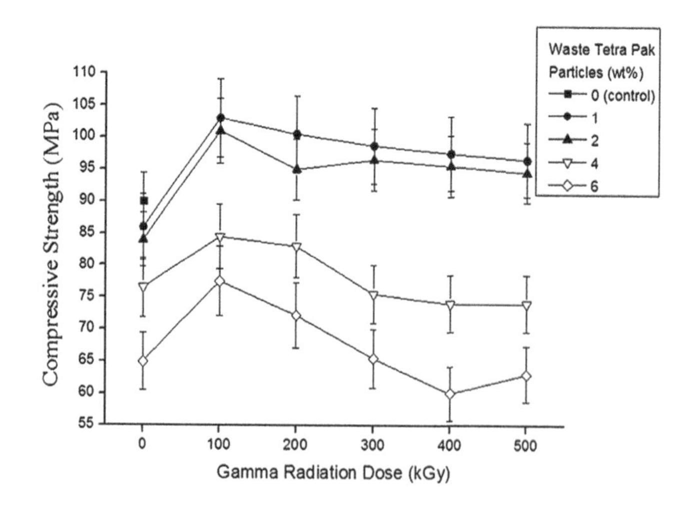

FIGURE 9.4 Variation of the compressive strength of polyester-based PC in terms of gamma irradiation dose and particles concentration [13]

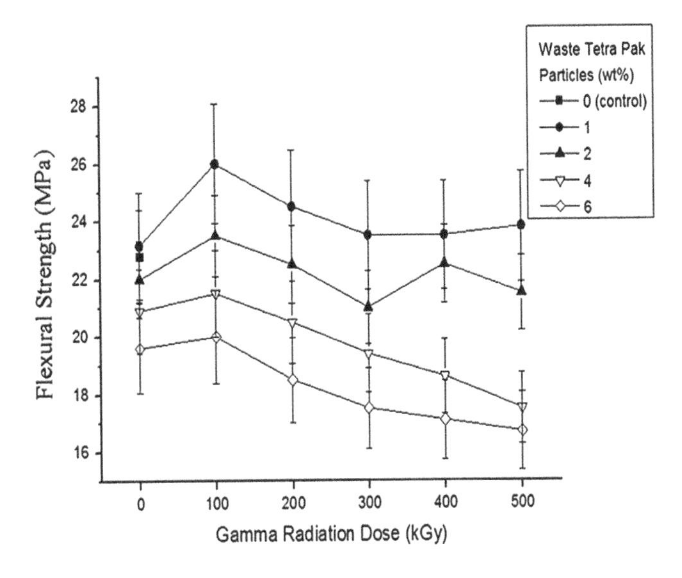

FIGURE 9.5 Variation of the flexural strength of polyester-based PC in terms of gamma irradiation dose and particles concentration [13]

consumption and exorbitant production cost of porcelain, efforts were made to replace PC as an electrical insulator. PCs post-insulation has a longer leakage or creep distance than porcelain insulators with the same height because of their structure that leads to a lower electrical stress. It should be noted that they perform better than porcelain at high voltages. Also, due to the lack of metallic end-caps in PC insulation system, they are lighter when compared to the porcelain insulation system [15]. Another advantage of PC over porcelain is its lower cost [16], and much easier manufacturing process, which can be done even at room temperature. Due to the remarkable chemical resistance of PCs against chemicals, they show more stability than porcelain insulation against acid rain [17]. PC insulators can be easily reinforced with various fibers to obtain the required strength. They also can be implemented in underground insulation systems in electrical power transmission, because of significant mechanical, chemical, and thermal properties [18]. Insulators made of PC have much better stability than porcelain against earthquake damage due to their high impact resistance and vibration damping [19]. PCs can be used as a high voltage (up to 138 kV) insulation material [20]. In a study on the long-term durability of electrical insulation of PC in different parts of the world with various weather conditions, significant performance of these insulators has been reported [21].

Based on the mentioned advantages, PCs are considered as a suitable alternative to electrical insulation such as electrical porcelain in switches, post-insulators for power systems and telecommunications, fuse cut-outs, transformers, capacitors, electrical energy sector, especially in underground applications, and so on [15]. Thus, it is important to evaluate their dielectric properties. However, only a few studies were performed on the improvement of insulation properties of PCs.

PC demonstrates approximately remarkable resistance to high voltage arc and the time to arc extinction is more than 300 seconds due to a relatively low polymer content in their structures. Implementing a higher amount of filler increases resistance to surface discharges, and the PC can be utilized in medium polluted areas [22]. Also, the dry arc resistance of PC strongly depends on filler content [11].

Electrical resistance, a parameter to evaluate the ion passed through the concrete, can be used as a criterion for corrosion probability. As can be seen in Figure 9.6, increasing epoxy content from 10 wt% (LWPC10) to 16 wt% (LWPC16) in a lightweight PC increases the electrical resistance due to the reduction of the porosity [23].

The conductivity of PCs can be reduced by combining some aggregates and fillers. Introducing chipped and crumb rubber as aggregates into epoxy PC significantly decreases the electrical resistance, as can be seen in Figure 9.7. In addition to types of implemented aggregates, the reduction can be attributed to the generated voids and porosity [24].

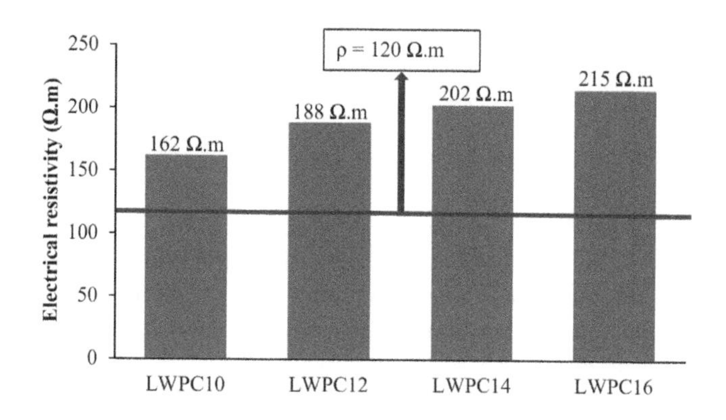

FIGURE 9.6 The electrical resistance of epoxy PC with different resin contents [23]

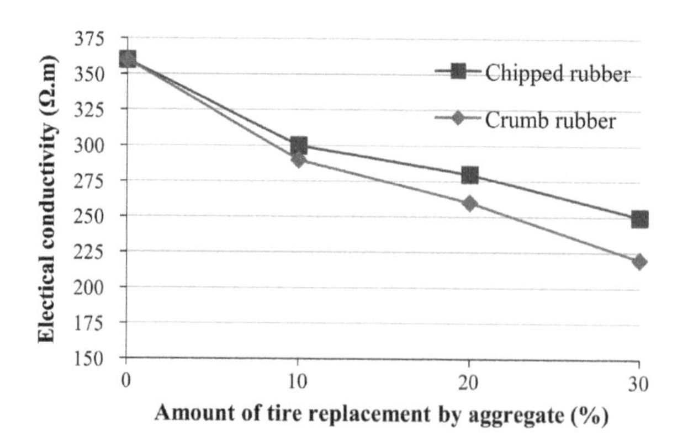

FIGURE 9.7 Variation of electrical conductivity of epoxy PC with rubber aggregates content (%) [24]

REFERENCES

[1] N.J. Jin, K.S. Yeon, S.H. Min, J. Yeon, Using the maturity method in predicting the compressive strength of vinyl ester polymer concrete at an early age, *Adv. Mater. Sci. Eng.* 2017 (2017) 1–12. https://doi.org/10.1155/2017/4546732.

[2] W. Lokuge, T. Aravinthan, Effect of fly ash on the behaviour of polymer concrete with different types of resin, *Mater. Des.* 51 (2013) 175–181. https://doi.org/10.1016/j.matdes.2013.03.078.

[3] J.J. Sokołowska, Long-term compressive strength of polymer concrete-like composites with various fillers, *Materials (Basel).* 13 (2020) 1207. https://doi.org/10.3390/ma13051207.

[4] J.J. Sokołowska, Long-term investigation on the compressive strength of polymer concrete with fly ash, in: M.M.R. Taha (Ed.), *Int. Congr. Polym. Concr.*, Springer, Cham, Washington, DC, 2018: pp. 275–281. https://doi.org/https://doi.org/10.1007/978-3-319-78175-4_34.

[5] X. Li, D. Ma, Z. Lu, Y. Liu, Z. Jiang, L. Tang, Z. Liu, L. Zhou, Experimental study of hygrothermal and ultraviolet aging on the flexural performance of epoxy polymer mortar, *Acta Mech. Solida Sin.* 34 (2021) 539–549. https://doi.org/10.1007/s10338-021-00234-y.

[6] W. Ferdous, A. Manalo, H.S. Wong, R. Abousnina, O.S. Alajarmeh, Y. Zhuge, P. Schubel, Optimal design for epoxy polymer concrete based on mechanical properties and durability aspects, *Constr. Build. Mater.* 232 (2020) 117229. https://doi.org/10.1016/j.conbuildmat.2019.117229.

[7] A.T.M.M.C.S. Ribeiro, A.J.M. Ferreira, Effect of natural and artificial weathering on the long-term flexural performance of polymer mortars, *Mech. Compos. Mater.* 45 (2009) 515–526. https://doi.org/10.1007/s11029-009-9104-7.

[8] K. Chung, Y. Hong, Weathering properties of elastic rubber concrete comprising waste tire solution, *Polym. Eng. Sci.* 49 (2009) 794–798. https://doi.org/10.1002/pen.

[9] R. Gri, A. Ball, An assessment of the properties and degradation behaviour of glass-fibre-reinforced polyester polymer concrete, *Compos. Sci. Technol.* 60 (2000) 2747–2753.

[10] W. Ferdous, A. Manalo, T. Aravinthan, G. Van Erp, Design of epoxy resin based polymer concrete matrix for composite railway sleeper, in: *23rd Australas. Conf. Mech. Struct. Mater., Southern Cross University ePublications*, Byron Bay, Australia, 2014: pp. 137–142.

[11] L.E. Schmidt, A. Krivda, C.H. Ho, M. Portaluppi, Polymer concrete outdoor insulation – experience from laboratory and demonstrator testing, in: *2010 Annu. Rep. Conf. Electr. Insul. Dielectr. Phenom.*, IEEE, West Lafayette, IN, 2010: pp. 1–3. https://doi.org/10.1109/CEIDP.2010.5723942.

[12] G. Martı, L.F. Giraldo, B.L. Lo, Effects of g radiation on fiber-reinforced polymer concrete, *Polym. Compos.* 29 (2008) 1244–1251. https://doi.org/10.1002/pc.

[13] M. Martínez-lópez, G. Martínez-barrera, C. Barrera-díaz, F. Ureña-núñez, Waste Tetra Pak particles from beverage containers as reinforcements in polymer mortar : Effect of gamma irradiation as an interfacial coupling factor, *Constr. Build. Mater.* 121 (2016) 1–8. https://doi.org/10.1016/j.conbuildmat.2016.05.153.

[14] R.E. Izzaty, B. Astuti, N. Cholimah, New concepts in polymer concrete insulation, in: *Second Int. Conf. Prop. Appl. Dielectr. Mater.*, IEEE, Beijing, China, 1988: pp. 5–24. https://doi.org/10.1109/ICPADM.1988.38392.

[15] M. Gunasekaran, Polymer concrete high voltage insulation: A versatile material for the 90's and beyond, in: *Proc. 21st Electr. Electron. Insul. Conf. Electr. Manuf. Coil Wind.*, IEEE, Chicago, IL, 1993: pp. 649–653. https://doi.org/10.1109/eeic.1993.631304.

[16] M. Gunasekaran, Polymer concrete: A viable low-cost material for innovative power systems, in: *Proc. 5th Int. Conf. Prop. Appl. Dielectr. Mater.*, IEEE, Seoul, South Korea, 1997: pp. 770–773. https://doi.org/10.1109/icpadm.1997.616549.

[17] R.E. Izzaty, B. Astuti, N. Cholimah, Polymer concrete high voltage insulation: A decade of progress, in: *1985 EIC 17th Electr. Insul. Conf.*, IEEE, Boston MA, 1985: pp. 5–24. https://doi.org/10.1109/EIC.1985.7458582.
[18] M. Gunasekaran, Polymer concrete: A versatile, low-cost material for Asian electrical infrastructure systems, in: *Conf. Rec. IEEE Int. Symp. Electr. Insul.*, IEEE, Anaheim, CA, 2000: pp. 356–361. https://doi.org/10.1109/elinsl.2000.845525.
[19] M. Gunasekaran, Basic forensic analysis of polymer concrete high voltage insulation in various applications, in: *2020 IEEE Electr. Insul. Conf. EIC 2020*, IEEE, Knoxville, TN, 2020: pp. 90–93. https://doi.org/10.1109/EIC47619.2020.9158668.
[20] M. Gunasekaran, Lightweight partially nano-particled polymer concrete: A new concept for electrical insulation, in: *2007 Electr. Insul. Conf. Electr. Manuf. Expo, EEIC 2007*, 2007: pp. 172–174. https://doi.org/10.1109/EEIC.2007.4562613.
[21] M. Gunasekaran, World-wide long-term outdoor performance of polymer concrete insulation, in: *Proc. IEEE Int. Conf. Prop. Appl. Dielectr. Mater.*, IEEE, Brisbane, QLD, Australia, 1994: pp. 515–518. https://doi.org/10.1109/icpadm.1994.414060.
[22] K.L. Chrzan, M. Skoczylas, Performance of polymer concrete insulators under light pollution, in: *XV Int. Symp. High Volt. Eng.*, Ljubljana, Slovenia, 2014: pp. 1–4.
[23] F. Heidarnezhad, K. Jafari, T. Ozbakkaloglu, Effect of polymer content and temperature on mechanical properties of lightweight polymer concrete, *Constr. Build. Mater.* 260 (2020) 119853. https://doi.org/10.1016/j.conbuildmat.2020.119853.
[24] K. Jafari, V. Toufigh, Experimental and analytical evaluation of rubberized polymer concrete, *Constr. Build. Mater.* 155 (2017) 495–510. https://doi.org/10.1016/j.conbuildmat.2017.08.097.

10 Summary and Outlook

Abstract

Altogether, given the unique properties, polymer concrete (PC) can be considered as an advanced construction material. In this book, we focused on the PC structures, materials (resin, micro, and nanofiller, aggregates, fibers), preparation methods, and properties.

The main research progress over the last two decades can be summarized as follows:

1. Implementing different materials (resins, aggregates, fillers, etc.) in polymer concrete (PC) structure to improve and optimize particular characteristics such as physical (weight, water absorption, density, porosity, etc.), mechanical (compressive, flexural, tensile, fracture, impact, etc.), thermal (coefficient of thermal expansion (CTE), conductivity, elevated temperature, and freeze–thaw cycles resistance, etc.), and chemical (resistance to aggressive media) properties.
2. Implementing fiber reinforcement strategy in PC and incorporating different natural and synthesis fibers to improve strength and durability.
3. Applying nanotechnology in PC and incorporating different nanomaterials for PC reinforcement.
4. Trying to reduce the cost of PC by applying locally available materials and also recycled resin and aggregates.
5. Applying different optimization methods for achieving maximum bulk density, reducing the number of tests, and obtaining the optimized PC with the best possible properties.
6. Preparing a PC with a specific characterization for a particular application.

Based on experience and research, the strategy of selecting the appropriate PC can be summarized as a block diagram in Figure 10.1.

In the first step, depending on the application and expectations of the PC product, the most important properties such as physical, mechanical, chemical, and thermal properties should be identified and the required magnitudes of each quantity should be determined. For example, to use a PC product in places that are exposed to high temperatures, the maximum temperature and time of exposure must be specified. The product may also be subjected to mechanical loading or chemical attacks at the same time. Therefore, all the expected properties of the product must be considered.

In the second step, using the knowledge of the properties of materials, resins, aggregates, and other fillers are selected to achieve the expected characteristics.

In the third step, after fabricating the specimens according to the preparation method described in Chapter 3, various tests are performed based on standards listed in Chapter 4. Afterward, the values of the studied properties are evaluated

DOI: 10.1201/9781003326311-10

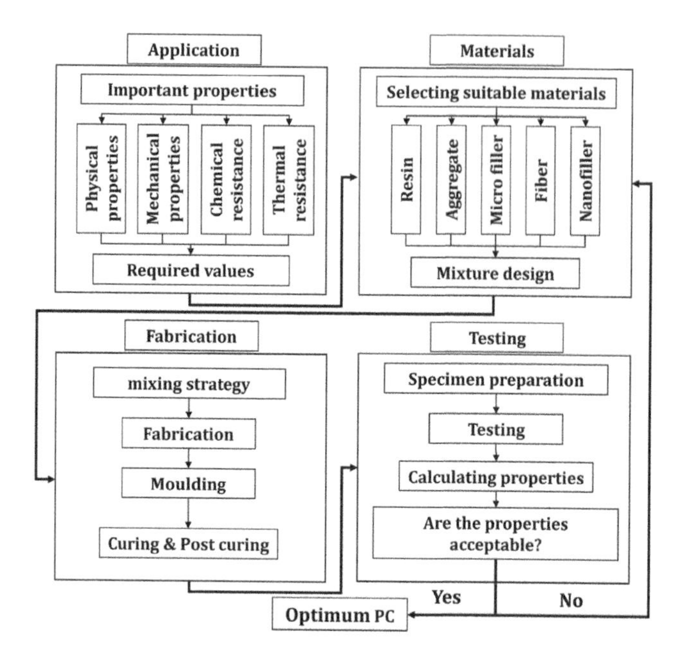

FIGURE 10.1 Schematic diagram of fabricating polymer concrete for a specific application

and analyzed. At this stage, the obtained experimental results are compared with the expected product values and if the specimens meet the expectations, the PC composition is selected as the appropriate composition. Otherwise, it is necessary to return to the second step and change the composition of the materials. At this stage, the properties of the samples can be improved to the expected properties by using additives such as fibers or nanoparticles.

The use of PC can't be overestimated, and the future of this technology is still open. PCs are considered sustainable materials because as a repair material, they increase the life of structures, and can save natural resources. Although considerable progress has been made in improving the PC properties, there is still a lot of studies that need to be performed. Here we list a few directions that might be worthwhile for exploration:

1. Investigating different properties of PC such as dynamic properties (i.e., damping and vibration properties, impact strength, etc.), fracture mechanism, and so on.
2. Use of nanotechnology to fabricate polymer nanocomposite concrete with excellent properties.
3. Study and fabrication of new PC generation with remarkable mechanical and endurance performances like self-healing, self-leveling, and self-sensing PCs.
4. Use of foam technology in PC to produce ultra-lightweight foam PC.
5. Investigating the applicability of PC for use in 3D printers.

6. Using molecular dynamics (MD) approach for simulation of the chemical interactions in PCs to design the mixture and predict the different properties of PC.
7. The use of different methods based on machine learning methods to optimize the mixture design and to predict the properties of PC.

Although most research to determine the properties of PCs has been based on experimental methods, recently attempts have been made to use mathematical methods [1], finite element analysis (FEA) [2], and molecular dynamics (MD) [3], as well as artificial intelligence [4,5] to determine and prediction of properties. Although experimental research provides far more accurate results, it requires more time and money. Applying theoretical and analytical methods to optimize the composition, as well as estimating and predicting the properties of these materials, can reduce manufacturing and testing costs to achieve the final product. Therefore, it is expected that theoretical and computational methods such as FEA and MD as well as machine learning-based methods will enter the field of PC more than before to enable the achievement of optimal formulation of these materials.

REFERENCES

[1] L. Kirianova, The fractional derivative type identification for the modelling deformation and strength characteristics of polymer concrete, *IOP Conf. Ser. Mater. Sci. Eng.* 1030 (2021). https://doi.org/10.1088/1757-899X/1030/1/012094.
[2] M. Troncossi, G. Canella, N. Vincenzi, Identification of polymer concrete damping properties, *J. Mech. Eng. Sci.* (2020) 1–10. https://doi.org/10.1177/0954406220949587.
[3] R. Bedi, S. Sharma, Y.K. Sonwani, Prediction of mechanical properties of epoxy concrete using molecular dynamics simulation, *Compos. Mech. Comput. Appl. Int. J.* 12 (2021) 25–39. https://doi.org/10.1615/CompMechComputApplIntJ.2021036343.
[4] B. Marinela, D. Rodica-Mariana, H. Maria, Using neural networks for prediction of properties of polymer concrete with fly ash, *J. Mater. Civ. Eng.* 24 (2012) 523–528. https://doi.org/10.1061/(ASCE)MT.1943-5533.0000413.
[5] R. Diaconescu, M. Barbuta, M. Harja, Prediction of properties of polymer concrete composite with tire rubber using neural networks, *Mater. Sci. Eng. B.* 178 (2013) 1259–1267. https://doi.org/10.1016/j.mseb.2013.01.014.

Index

Note: **Bold** page numbers refer to tables, *italic* page numbers refer to figures.